# The Dead Universe Theory

# The Dead Universe Theory

J. ALMEIDA

RESOURCE *Publications* · Eugene, Oregon

THE DEAD UNIVERSE THEORY

Resource Publications
An Imprint of Wipf and Stock Publishers
199 W. 8th Ave., Suite 3
Eugene, OR 97401

www.wipfandstock.com

PAPERBACK ISBN: 979-8-3852-2451-7
HARDCOVER ISBN: 979-8-3852-2452-4
EBOOK ISBN: 979-8-3852-2453-1

VERSION NUMBER 07/26/24

# Contents

# Acknowledgments

I WOULD LIKE TO express my heartfelt gratitude to my daughter, Sophia Milla, whose unwavering support and inspiration were fundamental motivations for undertaking this endeavor. I am also deeply grateful to my wife, Regiane Milla, for her steadfast encouragement and unwavering belief in me throughout the process.

Furthermore, I extend my sincere thanks to friends who always motivated me to continue writing, and to Eduardo Rodrigues and Shinoki, who challenged me to persist in writing this work after reviewing the initial chapter. This challenge was crucial to my decision to continue. With immense gratitude, I acknowledge their invaluable contribution to this project.

This work is the result of a series of lectures and presentations given in 2016 and 2017, which are available on video. It has been refined and revised based on the content of these videos.

# Chapter I

## The Creation of the Universe by Its End

THE DISCUSSIONS IN THIS book bring to the forefront the idea of a creator God responsible for the creation of the universe, presented as the first premise. In the second premise, the arguments are structured scientifically, based on cosmological models that do not rely on supernatural intervention to explain the origin of the universe. By eliminating the first premise, the origin of the universe can be described by an equation that offers a better explanation than the big bang theory. If we accept the first premise, the scientific conclusions of the work remain valid, as they are based on scientific foundations independent of the existence of a creator. The theory proposed in this work, called the "dead universe theory," provides a perspective on how the universe might have arisen without divine intervention but is also compatible with the idea of a universe originated by a creating force.

A new scientific theory for the creation and emergence of the universe will be proposed in the second chapter: the theory of the dead universe. This theory will be accompanied by a new vision of the expansion of the universe, grounded in empirical evidence confirmed by existing scientific data, presented in a new light. On the other hand, the events proposed by various theories for the end of our universe will be questioned in light of the new theory of the origins of the cosmos. However, a theological response will be offered to creationists, through a proposal of creation *ex nihilo* (universe that arose from nothing), deeper than those developed by biblical creationists until today.

The theory of the dead universe will take into account the various theories that present different perspectives on the origin of the universe, recognizing that many of them do without a Creator God, while this new cosmological model contemplates the probabilities of a universe created by God. This theory envisions a more comprehensive explanation of the origins of the universe and its continuous distancing process from galaxies, in accordance with the Hubble laws.

In this treatise, I will discuss the fascinating hypotheses I have developed to provide the reader with a simplified understanding surrounding the origin of the cosmos, starting with the concept of the great "empty womb" of the dead universe, and further exploring the hypothesis of the "galactic decrepitude" of the dead universe. These theories are conceived to shed light on the mysterious beginnings of our universe.

The argument that faith opposes scientific progress is unfounded and refuted by the historical record of the universities founded by Christians. The Christian effort to advance scientific knowledge, promoted by these institutions, demonstrates that Christianity was largely responsible for the scientific development of humanity.

We must put an end to this futile debate, sometimes fueled by individuals lacking knowledge of the history of academic development and the scientific progress stemming from these institutions.

The world's most influential universities have their roots in devout Christian foundations, a fact often overlooked by some academics and atheists who underestimate Christianity's contribution. These institutions emerged from an educational paradigm that values both faith and scientific inquiry. The integration of Christian principles with intellectual pursuits has fostered an academic environment that rigorously explores profound questions of existence and scientific understanding.

Within the halls of these universities, many renowned scientists and Nobel laureates have pursued groundbreaking research, unraveling the mysteries of the universe. This underscores the harmonious relationship between faith and reason, demonstrating that religious convictions can coexist with scientific exploration.

In the twenty-first century, it is imperative to avoid intellectual dishonesty by turning the debate into a confrontation between science and religion. Arguing against religions using science as a weapon is counterproductive and does not reflect the true essence of the quest for knowledge. I believe that to disparage the contributions of religions is to overlook the merits of this gift from the Creator to humankind. To disparage science would be akin to a person mocking the very entity that has bestowed upon them benefits they cannot live without in order to find happiness.

A man of elevated spirit compared to others may adopt a position of faith in a particular belief, but above all, he will be guided by the teachings of his master: "Love your neighbor as yourself, and you will fulfill the entire law." Therefore, this work is not a defense of any particular religion, but above all, proposes belief in a singular Creator God and an approach of respect and appreciation for science and its contributions to humanity.

The rationale behind a direct engagement with Christianity and Jesus is to underscore the presence of intellectual integrity within these philosophical discussions. It emphasizes the importance of not forsaking truth for the sake of self-interest, as has been observed in certain astrophysicists and scientists who advocate for the demise of Christianity. This religion, historically, has been a paramount contributor to the advancement of global academic development.

We should elevate the debate to a higher intellectual level, recognizing that both science and religion have their place and value in our lives. Instead of seeking conflict, we should strive for understanding and mutual respect among different forms of knowledge. This will allow us to fully appreciate the richness of the human experience and lead us to a more constructive and enriching dialogue.

It's worth noting that the first fifty universities in Europe and the United States, many of which consistently rank among the world's best, trace their origins back to Christianity. This historical fact highlights the significant role Christianity has played in shaping the landscape of higher education and scientific inquiry.

Spanning the United States and the United Kingdom, a group of prestigious universities founded by Protestant Christians has left an indelible mark on the academic fabric, collectively accumulating a remarkable number of Nobel Prizes: Harvard University (47), Massachusetts Institute of Technology (78), University of Cambridge (89), California Institute of Technology (33), University of Oxford (48), Stanford University (27), Princeton University (36), University of Chicago (87), Yale University (49), Columbia University (82), University of California, Berkeley (51), and Imperial College London (15), each contributing to a legacy of excellence and innovation in their quest for knowledge and discovery.

Over approximately twenty-one years, I have studied the existing theories and scientific data on the observable universe and interpreted them in light of the earliest texts of the Torah, seeking to glimpse the possibilities of some correlation between these events. Therefore, this work was not written for atheists or believers, and I do not aim to prove that one theory or another is correct, but merely to raise essential questions that I believe deserve the attention of astrophysics and theology, whether you are a skeptical scientist or a confessing Christian.

The theory of the Big Rip is a cosmological proposal that suggests an extreme fate for the universe. According to this theory, the expansion of the universe is occurring at an accelerated pace due to the so-called dark energy. Over time, this dark energy causes even greater acceleration in expansion, leading to the eventual tearing apart of all matter in the universe, including atoms, galaxies, and even fundamental particles. This catastrophic event would be known as the "Big Rip," where the very fabric of space-time irreversibly disintegrates. This theory is one of several hypotheses about the ultimate fate of the universe and is the subject of study and debate in modern cosmology.

The eschaton in creation, a term derived from the Greek ἔσχατον, not only denotes the end of the creative process but also suggests that God began his work by the end, revealing his sovereignty over time to fulfill his eternal purpose from the beginning. The teaching of creation is of great importance to Christianity. In this theological treatise, we will delve into the theme of the eschaton in creation, using Gen 1:1 as a basis, where the Greek term "end" is used. Additionally, we will seek support from various biblical texts to enrich the understanding of the most important event in human history.

The way God manifests himself in Gen 1:1 is sufficient to understand that he is an almighty God, capable of creating the universe and everything in it, both the visible and the invisible, including the possibilities of multiverses and multiple realities. From my perspective, he begins to diminish himself when he becomes involved in the project of creating man, considered the most complex and ambiguous of his creations. As he relates to humanity, he seems like a small, fragile God, demonstrating feelings and affections that do not seem appropriate for an almighty God. Throughout the sacred canon, he seems to diminish himself more and more as he becomes involved in the daily affairs of men. Even worse is when he demonstrates his infinite power in events like the Exodus of the people of Israel from Egypt. As this people advance in territorial conquests, this almighty God becomes equal in size to the existing pagan deities up to that point, as he becomes involved in territorial issues and promises to clothe his chosen kings with power, something common in the beliefs of monarchs who did not know this God who diminished himself to the point of asking for the construction of the Ark of the Covenant to relate to men. Not satisfied with diminishing himself so much, Yahweh takes on human form in the incarnation of Jesus Christ to demonstrate his love for humanity, becoming like us so that we do not spend eternity with a strange and unknown God like the gods of other religions. "Let this mind be in you, which was also in Christ Jesus: Who, being in the form of God, thought it not robbery to be equal with God: But made himself of no reputation, and took upon him the form of a servant, and was made in the likeness of men: And being found in fashion as a man, he humbled himself, and became obedient unto death, even the death of the cross. Wherefore God also hath highly exalted him and given him a name which is above every name" (Phil 2:5–9).

All other gods, if they exist, must envy humans, since they do not know what it is to experience their own death. They are trapped in eternity, gods like Yandroth: the supreme scientist of his universe, pitting the combination of his technology and knowledge of sorcery against Doctor Strange's magic; Zom: the most powerful demon in existence, beyond even the ability of eternity itself to defeat. A God who lacks the power to become equal to us cannot be a good God. A God who creates an extensive code of laws to be followed and doesn't even have the audacity to descend to earth to fulfill his own laws, imposing them on humans as if they were his slaves. These gods are detestable, for they kill but do not want to die; they cause hunger and thirst but do not know how to feel them. They sit on thrones while watching human miseries and delighting in their feasts in eternity. They have never been spat upon, never felt fear, insomnia, or loneliness. What do these gods share of human life? Can they feel the pains of a just man condemned to the harsh penalties of the cross under the Roman Empire's death sentence? Did they feel the pain of betrayal, the exposure of their shames, have their beards plucked from their faces? Did they subject themselves to a trial in human courts?

The God who became equal to us is Jesus Christ. According to Christian theology, Jesus, being God in human form, came to earth to live among human beings, experiencing the same pains, joys, temptations, and challenges we face. He fully shared in our humanity, living an earthly life like ours, so that he could understand our needs and offer us salvation and redemption. This is a fundamental belief of Christianity, that Jesus is both fully God and fully human.

He, Christ, gives purpose to the universe we know and all unknown dimensions, making us inexcusable before God. It is indecent to say that there is no purpose in the universe and, at the same time, consider oneself an intelligent being. An even greater absurdity is to contemplate the existence of a purposeful God and lead a life without purpose. The simple definition of purpose is the reason why something exists, and if it exists, someone desired it intentionally. And if there is purpose in created things, there can be no absence of a creator. Otherwise, the universe and all life within it would have no purpose. What motivating challenge is there for a universe and a life without purpose? It is impossible to find reason for something without purpose; if there is no reason, then it does not need to be explained. The improbable is the coup de grâce to science. This is painful for all of us. It is also frustrating because it would nullify the importance of investigative function once and for all. We know that many things are beyond the reach of science, and among them is the blind assertion of the purposelessness of the universe.

Every day, the sciences of various branches delve into many subjects related to their knowledge. For now, nothing in the fields of chemical research, astrophysics, quantum mechanics, geology, astronomy, and cosmology provides direct evidence for, or against, this divine purpose in the universe. We have no logical permission to corroborate anything about this senseless void. In fact, science may never be able to prove this false assertion that such purpose does not exist. Carl Sagan, one of the greatest scientific authorities on the causes of the universe, once said: "There are many hypotheses in science that are wrong. That's perfectly fine, they are the opening to find the ones that are right, and the absence of evidence is not evidence of absence."[1] What will leave us more perplexed and haunted in this unknown cosmos? Is all of this the work of a creating God? And if it is his work, does it imply serious responsibilities in light of this truth? Is our intelligence in the universe, as it is, not clear evidence of purpose?

I believe that after the advent of original sin, our lives would no longer be interesting without the freedom granted to us. "He was led like

1. This quote is frequently attributed to Carl Sagan and reflects his scientific perspective. However, a detailed review of his published works, including *Cosmos*, did not reveal this exact phrase. Therefore, it should be considered a paraphrasing or synthesis of Sagan's views on science, rather than a direct quote.

a lamb to the slaughter, and as a sheep before its shearers is silent, so he did not open his mouth. The next day John saw Jesus coming toward him and said, 'Look, the Lamb of God, who takes away the sin of the world!'" (John 1:29–36).

Be content, for life itself has purpose, for the greatest purpose of life lies in life itself here and now. Its essence is undoubted in the sacred Scriptures, for this reason, we know where it came from, but its purpose is inherent in itself. "He has made everything beautiful in its time. He has also set eternity in the human heart; yet no one can fathom what God has done from beginning to end" (Eccl 3:11).

Science cannot explain where life came from, but it also cannot assert that it exists without purpose. But it is a fact that it has an origin. Is there an origin without a source? We have two objections, either it originated from the universe itself, and in a second premise, it is a product of divine purpose. We know that thinking about it as divine purpose is more beneficial, but if we accept the neurosis that all this immense universe is purposeless, we are already fried in the fires of hell. If we are alone in this inert and meaningless retardation, the best solution will be to delude ourselves into horrors, for the vengeance of the universe's own nature has no mercy on us.

We are not self-existent beings; therefore, scientific theories that portray humanity as self-sufficient entities capable of solving all the world's problems jeopardize our very existence. We transcend what science perceives of us; the future of the human race cannot solely be found within laboratories. We must not become masters of our own destruction; solutions to prolong humanity's existence cannot be found solely through space observations. The loss of meaning in life has become the prognosis of death for humanity.

Nihilism, associated with Friedrich Nietzsche, is a philosophical current that elicits divergent reactions. Some worship it, others repudiate it. Its hallmark is the "death of God," undermining concepts of morality and existential meaning.

This philosophy, built upon a radical refusal of once-sacred values, reveals itself as a seed of deception. By challenging Christian morality and denying the existence of God, nihilism proves to be a futile resignation, devoid of potency.

I perceive how nihilism promotes the disregard of meaning in favor of a morality-free freedom. This implicit denial is an expression of human discontent, leading to existential emptiness.

If we consider belief in God as a moral order, defending nothingness as a solution is unsustainable. Nihilism arises amidst moral frustration, becoming a refuge for those lacking an absolute truth.

The nihilistic individual feels trapped, losing the sense of rightness amidst the void. There is no entirely secular morality, and "absurdity" becomes necessary to maintain a consumerism-based psychology of happiness.

With the "death of God," nothingness becomes the norm. However, we perceive that there is no order in the "patterns of nothingness" to build a new world. Nihilism represents a potential for postmodern psychological illness, leading to the loss of faith in Christianity, presented as an object devoid of meaning.

The teaching about creation is fundamental to our Christian faith; belief in God is not a delusion but a response to the void of nothingness proposed by nihilism. It is a central doctrine in Scripture.

Sometimes, we may have a profound understanding of various doctrines, such as the kingdom of God, and extensive knowledge about Jesus. However, it is evident that when someone lacks a deep understanding of the doctrine of creation, they may face crises, especially when confronted by atheists or university professors. Some educators present ideas that strongly contradict the creationist doctrine. For example, the belief in a six-thousand-year-old Earth, originating from the seventeenth-century archbishop James Ussher, lacks scientific basis, although it may have theological foundations. As theologians or scientists, it is our duty to instruct people based on solid evidence and grounded truths, rather than propagating unfounded theories. We must understand that the biblical narrative does not aim to provide an exact count of years but rather a poetic description of God's creative process.

Often, some Christians are criticized for believing that the Earth is six thousand years old, as if that were synonymous with lack of intelligence. Within Christianity and Judaism, themselves, there are various schools of thought that diverge on the dates of the universe's creation. This is a matter of faith in the power of God; the Earth may simply have been created aged by God himself to confound the wise. However, scientifically, we know that the Earth is around 4.5 billion years old. It is crucial to demystify this mistaken notion that Christians are ignorant for thinking otherwise and that there is a form of thinking that supersedes the other, but in fact, they do not even refute each other. The aim of this enlightening work, which sheds light on biblical texts, is to help young

Christians face challenges in university. We acknowledge that both points of view, both faith-based and scientific, have their value and can complement each other. While faith allows us to consider the possibility of a creation aged by God, science offers us valuable insights into the age or another possibility of age for the Earth. Both knowledge can coexist and enrich our understanding of the world around us.

This work is for all of us, Christians and atheists alike, to reflect on creation, recognizing the sovereignty and power of God over all things. He is the Creator who brought into existence everything we know and what we do not know. As we study creation, we are led to contemplate the greatness and wisdom of God, which is beyond our human understanding.

We live in a world of chance, where truth blends with lie, and what was once considered false has now become our truth. Christianity, built upon a moral perspective, has collapsed, and now it is imperative to rethink the paths to be taken for a new era of thinkers in a sick contemporary world. We can no longer be content with seeking answers only in the roots of faith to a world that has strayed from the teachings of biblical creationism.

In a world ravaged by pain that tears at human hearts, many fallacies intertwine in a false morality presented as a prognosis to heal the ailments of the human psyche. It is common to see a multitude of unfortunate souls seeking solace in therapeutic offices, a conception of mental health that has become commonplace in the profit machine of pharmaceutical-scientific capitalism.

We are reaping the fruits of our fantasies; we are being crushed by the materialistic world that has annihilated the noblest senses of existence, making room for nihilism to thrive. Christianity, in its ritualistic form, has often become an answer to the world's pains but has ended up being confused with distorted messages from false prophets of superficial healing. The circumstances that afflict some become the same that elevate others in the philosophical debate about truth. For many, the functionality of lies is worth more as a momentary relief of conscience than seeking the truth.

We have become slaves to our own freedom, forged by our own empty thoughts. Many intellectuals of liberal theology in the United States lack a relevant practice of expressing the sense that celebrates awareness of the right to know the truth, under a spiritual perspective in Jesus Christ. Their lives are governed by the ironic "secular morality," which serves as an escape from conscience issues diverted from belief in God.

The adherents of placebo Christianity also deserve attention. These false moralists, distant from the liberation theology promoted by the resurrection of Christ, pride themselves on promises of Christianity merely as an expression of belief, without grounding in truth. This type of prefabricated Christianity, rooted in magical portions of the stimulus-producing brain, may even offer a sense of improvement through stiff rituals, but it superficially heals people of their wounds, feeding the capitalist selfishness that has seized neurotic souls.

On television, we often see false prophets promising solutions to financial problems, claiming to expel demons, all through a placebo Christianity that conveys the idea of peace and tranquility without true light. These prophets offer hope to those hungry for spiritual crumbs, charging a high price for the services offered. The prophet Jeremiah warned about these superficial healers: "They have healed superficially the wounds of my people, saying, 'Peace, peace,' when there is no peace."

The theological position of Jesus Christ presented God's grace as a cure for the pains caused by the moral law embedded in the human spirit. The early Christians understood Christian ethics as the liberating truth in the resurrection of Christ, living the values of self-denial for the love of Jesus Christ and his teachings. They conceptualized the meaning of life itself, not as a lifestyle governed by false morality as Nietzsche intended to accuse Christianity of being responsible for depriving life of its meaning in the world.

Before God created the cosmos, he already had a perfect plan in mind. Creation was not a random event but part of God's eternal plan for humanity's redemption in Christ. The Son of God was present at creation and is the center of this redemptive plan known from the beginning. We must reject erroneous interpretations that diminish Christ's divinity or question his pre-existence.

If we want to understand a God who truly understands us, who walked in our shoes and felt the weight of our humanity, we need a God who has been human. A God who not only observed our existence from afar but experienced every aspect of it firsthand.

A God who knew the pain of being slapped, of going hungry and thirsty, of feeling naked and despised, of being spat on and humiliated. A God who faced the cold of the night and the scorching heat of the day, who knew the anguish of persecutions and the loneliness of human temptations.

Because, after all, a God who never served and felt all of this cannot truly understand us. If Jesus was not truly God incarnate, then we would be destined to spend eternity with a being who never truly cared about being human, a God strange to us.

It doesn't make sense to imagine an eternity alongside such a being, no matter how all-powerful. Therefore, the idea of Jesus as God incarnate is essential for our understanding of God and for our hope of a meaningful eternity close to our own being.

As we study texts like Isa 49:9–10 and Col 1:16, we understand that God is the creator of all things, visible and invisible. Everything was created by him and for him, for his glory. Even spiritual realities, like angels and the devil, are part of this divine plan and contribute to the manifestation of God's glory.

Therefore, as we discuss creation and eschatology, we must do so based on Scripture and the truth revealed by God. We must seek a deeper understanding of creation as part of God's redemptive plan and recognize the centrality of Christ in this plan.

As a theologian, it is essential to address the creation narrative without omitting the context prior to the creation of the heavens and the earth by God. This approach is fundamental to understanding the role of Jesus, who was present in creation before the foundation of the world. When examining Scripture, it is observed that some religious groups face theological conflicts, especially related to the doctrine of the Trinity, due to the complexity of Hebrew. For example, the word "Elohim," which is plural in Hebrew, also refers to God as a title. This issue needs to be explored in depth.

Upon examining Gen 1:1, which states, "In the beginning, God created the heavens and the earth," it is observed that this passage generates controversy among certain groups, including an emerging faction that advocates for an alternative interpretation of the term "Elohim." For them, this interpretation suggests that Jesus was the first creation of God, a proposition considered heretical without theological or biblical foundation. According to the Scriptures, Jesus was not created, but preexisted before the foundation of the world, manifesting as God incarnate. He is the Son of God, who took human form, as consistently affirmed in the Christian tradition.

The difficulty in understanding Hebrew and the mistaken interpretation of Gen 1:1 can lead to theological embarrassment. It is crucial to conduct an in-depth study of this period prior to creation, exploring verses like Isa 49:9–10, which highlight God's sovereignty and designs from eternity. God operates in ways incomprehensible to the human mind, and his thinking transcends our understanding, as evidenced in Isaiah.

Considering that God "starts things at the end," it is important to understand that he already has a complete plan for each individual. Just as Jesus had a predetermined journey culminating in his death and resurrection, each person is embedded in a historical plan established by God. This understanding is crucial to avoid erroneous and heretical interpretations about the nature of Jesus and the divine creative process.

We who believe in election firmly uphold this conviction. Scripture is clear in affirming that we were chosen in him before the foundation of the world and predestined in Jesus. Therefore, to God, our life is a finished project, while for us, it is a journey of faith and walking. God, in his wisdom, knows from the beginning the outcome of all things. As mentioned in Col 1:16, all things were created through him and for him.

When discussing creation, it is essential to recognize the distinction between the theological and scientific perspectives. While theology asserts that God is the creator of the heavens and the earth, science seeks to understand the natural processes and laws that govern the universe.

The apostle Paul, in writing to the Colossians, emphasizes that all things were created through Jesus and for him. This includes not only visible things, such as galaxies and subatomic particles, but also invisible things, such as angels and other spiritual realities.

In addressing creation, it's as if we were affirming that Steve Jobs created the MacBook. This is an unquestionable truth in terms of technology, just as in theology the assertion that God is the creator of all things. However, science has the role of investigating and uncovering the mysteries of creation, both visible and invisible.

The reference to visible and invisible things in creation highlights the greatness and transcendence of God, who is the author and sustainer of all creation. Even the devil was created for the glory of Jesus, and the name of Jesus was glorified in his victory over evil.

In the Gospel of John, Jesus is presented as the Word who was with God from the beginning and who is identified as God. This confirms his preexistence and active participation in creation, as he is the creative principle of all things. Although he is called the "firstborn of all creation," this does not imply that Jesus was created, but rather that he has preeminence over all creation, according to the sovereign will of God.

The lack of hermeneutical study by some people is evident, as they neglect the essential principles of interpretation. According to rabbinic laws, to establish a truth, seven verses or three similar verses are necessary as confirmation. However, in the New Testament, there are not seven verses that affirm that Jesus was created, or even a single verse with such a statement. The expression "firstborn of creation" refers to the fact that Jesus is the head, not a creature of God, as some may interpret erroneously due to lack of knowledge of the original languages, Greek and Hebrew.

When we examine Rev 21:6, we perceive the description of God as the Alpha and the Omega, the beginning and the end of all things. Implicit in this passage is the idea that Jesus is the exact manifestation of divine glory, for in him dwells all the fullness of God. He is the beginning and the end, the exact extension of all things, having even conquered death by rising on the third day.

Therefore, it cannot be denied that Jesus is God, for he conquered death and is the beginning and the end of all things, as declared in Revelation. He is the Alpha and the Omega, the very God incarnate. As stated in John 1:2, "In the beginning, he was the Word, and the Word was with God, and the Word was God." This truth is indisputable, both in Portuguese and Greek, not allowing distortions or mistaken interpretations. He was with God from the beginning and is God in his essence.

And Jesus declared himself to be the beginning and the end, existing before the foundation of the world, as evidenced in 1 Pet 1:20. His manifestation occurred out of love for humanity, as the same verse indicates. Thus, even before the beginning of creation, Jesus already existed as the beginning and the end, being the very incarnation of the word of God.

According to the Scriptures, God created all things through his word, and Jesus is that living word, the incarnate Torah, representing the Scriptures in their essence. He was the living agent who manifested creation, existing even before the beginning of the world. The promise in Revelation about the new creation reveals that God's work in creation is not yet complete, indicating a new heaven and a new earth.

Biblical theology upholds the belief in the recreation of the Earth, a transformation that will be enduring before God, as announced in Isa 65:17. This expectation of the new heavens and new earth, where righteousness dwells, underscores God's constant creative desire, demonstrating his sovereignty and his nature as the Supreme Creator. Thus, we understand that God does not create out of necessity or lack, but out of his grace and intrinsic desire to manifest his glory.

God does not need to be worshiped, as many may mistakenly think. His worship does not arise from a need or personal desire, but from the absolute sovereignty he holds over all things. His divine nature transcends any human need for approval or recognition. He is worthy of praise not out of his own need, but because of his unmatched greatness and majesty. Unlike human beings, he does not seek applause or depend on a platform to assert his supremacy.

In fact, God does not face any identity crisis. His essence as Creator is intrinsic to his divine nature. While human beings have a natural propensity to create, it is interesting to note that, unlike us, who create from what already exists, God has the unique ability to create from nothing, which highlights our likeness to him, as we are made in his image. On the other hand, the devil, as the father of lies, does not possess this creative ability. His only creation is lies, and his envy toward our ability to create is evident.

The devil can only invent lies. Lie is his creation, for he is recognized as the father of lies. Meanwhile, humans, despite all the problems they face, are capable of creating computers, cameras, cars, and airplanes from what already exists. This creative capacity reflects our condition as made in the image and likeness of Elohim. We were baptized with the spirit of Elohim.

Therefore, God has granted us creative ability. Within each of us are potentials that God has planted, such as businesses, projects, and lives that he desires to use to bring other lives into his presence. If people understood the potential seeded by the Spirit of God within them, they could overcome crises, depressions, anxieties, sorrows, and uncertainties. There is something extraordinary of God within each one, a capacity to generate and create. For some people, it is easier to start a business than to find a job, as they have the potential to become entrepreneurs. This depends only on their will, while getting a job nowadays can depend on various circumstances, including its existence, whereas the entrepreneur can create their own work.

Some people spend two years looking for a job and worry about creating an opportunity to, from that, be able to exercise their skill, their gift, their talent; perhaps they wouldn't be suffering so much. Normally, social concepts and ideologies try to inhibit the creative capacity that each one possesses, given by the Creator. He has empowered us. The devil cannot create anything, he will simply lie to you, he will say that you cannot, that this is not for you, that you do not have the ability. The family you were born into, your origins, the color of your skin, do not determine your limits. My brother, my sister, God has sown his capacity within you, he has given you the spirit for it.

You don't need to feel distressed and depressed when looking at this world that the Lord has created and placed in your hands, endowing you with abilities and talents to advance, both as a people of God and as a church. It is surprising to see so many people today, with so much potential, serving God, but also serving the devil and demons, when they could be using their gifts for the glory, praise, and honor of God. The Scriptures affirm that everything was created for him, for his glory, his honor, and his praise. Therefore, understand that you are much more than you imagine, for the Lord who created you thus declared it in the Scriptures.

Since 1945, many theologians have spent much time discussing creation based not only on Gen 1:1 but also on the first chapters of Genesis, up to chapter 4. Some argue that these first chapters of Genesis 1 to 4 actually constitute a confessional creed of the Jewish community. If this is true, it would greatly limit our understanding of creation, as it would be a view based solely on the experience of the Jewish community, which believes, for example, that God created the heavens and the earth, the animals and man in his image and likeness, just as Christians believe. We can also approach this line considering it as a confessional creed for Israel.

But in truth, the Scriptures make it very clear that Genesis, chapter 1, from verse 1 to 4, represents God's revelation to humanity. It is God's will for humanity, it is his desire for us. It is what God intends, what he wants for humanity. Therefore, if we do not dedicate ourselves to understanding the Old Testament with fear and trembling, we will not be able to reach the conclusions that God revealed for our lives through the Scriptures. The Old Testament is the beginning of God's revelation, where he reveals Himself as sovereign and creator.

In Gen 1:4, Elohim is not trying to prove his existence. He is not saying, "Hey, I am Yahweh, here I am." He does not present himself in that way. Instead, he is bringing his progressive revelation to humanity, through the community of Israel. We know that Genesis was written by Moses. Some argue that he collected much material from the traditions of his ancestors and included these traditions in Genesis, including chapter 1, verse 4. Moses brought all this oral tradition to include in Genesis.

In the city of Jerusalem, ancient Shalem, there possibly existed a tradition of Torah and Scripture study, and it is likely that this line of teaching was passed down from generation to generation. According to what we study in the Scriptures, there are mentions of the king of Shalem and the message of Shem's tents, suggesting that Shem was possibly the son of Noah and that he had a school to teach the Scriptures. Possibly portions of the Torah, in its oral tradition, before Moses, as we advance to a perspective that the Torah is simply eternal.

The revelation contained in chapters 1 to 4 of Genesis about creation reached Moses, and he included it in the Scriptures. Elohim also gave Moses direct revelation during the forty days he spent on the mountain, and Moses wrote this account, leaving the record we have today. When Elohim's will manifested for Yisrael to be the people of the promise and to reveal his covenant to humanity, this confessional creed was consolidated. Therefore, the Old Testament is Elohim's revelation to humanity, expressing his desire and purpose for all of us. It is what Elohim shows regarding humanity, his congregation, and his people Yisrael. This is the view we should have of the Old Testament. The Old Testament is something serious. Many people strain out a gnat but swallow a camel. In this work, we will study everything, we will talk about revival and other subjects. We have much to understand about the creation vision in the Torah. Because we need to be grounded in the Scriptures, and creationism is an essential doctrine, vital to our faith. We are in a completely secularized world. The doctrine of creation is totally essential to our faith. And we need to delve into this biblical doctrine.

The younger you are, the more you need to understand about creation. I, as well as my wife and daughter, are passionate about the book of Genesis and the book of John. There is a connection between the two books. Some theorists claim that the book of John is the Genesis of the New Testament. And John's declaration in chapter 1 is a reference to the beginning, to creation. Therefore, we must have respect and reverence for the Old Testament, which should not be neglected, ignored, or left unstudied. This needs to be changed, for we have a confessional creed. But Christian communities with weak leaders on theological issues of creation will have people full of doubts in the flock.

Even science, through astrophysics, affirms that the entire visible universe originated from a "small sphere" that exploded, giving rise to everything we see. Thus, many people are turning to other directions that even astrophysics and cosmology are not going in.

Some are trying to reconcile the theory of species evolution with the theory of divine creation regarding man and woman. However, they themselves cannot explain the origin of dinosaurs, for if the Earth is only six thousand years old, there would be no room for the existence of dinosaurs. On the other hand, following the theological cord view of this work, we solve the problem of the existence of dinosaurs and the age of the Earth.

If we affirm that dinosaurs did not exist, this can challenge our faith. On the other hand, if they existed and the Earth is just over six thousand years old, this suggests that humans lived with dinosaurs, but we will have serious problems with academic theories. However, there are no written records nor cave art representations of these animals by humans of that time. But if we consider that God's time is different from men's time and that the Earth is 4.5 billion years old, there is no problem with the existence of dinosaurs, for humans did not share the Earth with them. The Scriptures affirm that a day for God is equivalent to a thousand years, which suggests that the Earth can safely be 4.5 billion years old. If we do not study as the Scriptures teach us, we may encounter challenges to our faith. Verse 1 of Genesis is crucial in this aspect, as it establishes that in the beginning God created the heavens and the Earth, and that, to me, is like the creation of the iPhone by Steve Jobs. Before him creating the device, there was a previous creation that took much longer than the aspect of the finished device we know. This easily explains the theory of a 4.5 billion-year-old Earth. God's creation time is not the same as men's time; in fact, there is no time in eternity. Before Jobs showed the first iPhone to the world, other companies had created chips, boards, and other incredible innovations for the device to present itself as ready, but there was time and previous investments. This is the same idea behind Genesis. In the beginning, God created the heavens and the earth, and throughout the account, he explained the creation process in detail; no one can imagine the time required in this process. He said "let there be light," "let there be this," but someone at a spiritual maturity level does not worry about moving on to Genesis chapter 3, because the previous chapters explain the process of the entire creation; but verse 1:1 and a complete presentation of the "product" just as you have the final iPhone in your hands and are not interested in how the previous process was to have it as you have it. If we already understand that God created the heavens and the earth in Gen 1:1, and the Bible confirms this further, the problem is solved. The big issue is the lack of attention to what the text really means. Many people try to force interpretations of the Old Testament, but this can lead to problems, for the Old Testament is God's revelation. Reverence is needed when studying the Old Testament, for it is the word of God.

The word of God is for Christians, Jews, and humanity. Therefore, there is much lack of fear and respect regarding this. It is infallible. Someone asked me: "But is it infallible?" Because it is the word of God! Inspired! Can you explain to me what that means? God breathed upon his servants and prophets and inspired them to write. As it is inerrant. It is inerrant, how so? It is not about perfect words, without translation errors, missing verses, or adding a verse. Even if they were added before closing the sacred canon, it was the will of the Creator, but that is not what we are talking about. Inerrant not in that sense. In the sense that God gave the text, it is his will, everything he gave to his servants and prophets is his word, it is complete, it is perfect.

Now, if I decide to translate a Bible and insert what I want, then that becomes a human problem, not a divine one. Many resort to the argument of the original language, but, in reality, we do not have direct access to it. The oldest materials of the Old Testament, from the Dead Sea, date from periods later than those of Jesus, resulting in translations of translations. It is common for those whose faith is shaken to seek answers on the internet or listen to the opinions of skeptics, which can lead to a crisis of faith. However, it is important to remember that God is sovereign and watches over his word. Does God need human means to achieve his purposes? Although he provided us with the Bible to instruct and prepare us, it is he who rules over his word so that his designs are fulfilled. Thus, in this sense, the word of God is truly inerrant, inspired, and breathed by the Spirit. Scriptures are divinely inspired by the Spirit of God to guide, correct, and edify lives. Therefore, paying little attention to the Old Testament can result in challenges to our faith and spiritual understanding.

Certainly, when confronted by individuals who possess an innate ability for critical thinking, we are often led to question our own beliefs and doctrines. This can generate doubts and uncertainties, especially regarding the doctrine of creation. However, it is crucial to understand that this doctrine is fundamental to our faith and understanding of life. When reading passages that affirm that God created the heavens and the earth, we may be perplexed by the vastness of the universe, like the Milky Way with billions of stars and black holes. However, we must remember that God is eternal and infinite, not limited by the cosmos. He is described as sovereign over everything, and everything is in his hands. Therefore, even in the face of discoveries in astrophysics, we can trust in the truth of the Scriptures.

"Let there be stars! Let there be earth! And it happened." The Old Testament is God's revelation to humanity prior to the incarnation of Christ. He reaffirmed this revelation, referring to Genesis and Abraham. Jesus is our hermeneutical key. To determine if a book is inspired, we must ask if Jesus or the apostles cited it. As Jesus often mentioned the book of Genesis, it is unquestionable for the Christian faith. The person who does not accept creationism is estranged from healthy evangelical faith. If one does not believe that God made man in his image and likeness, this will be addressed later. The record of revelation was entrusted to Israel, but it was given for us to understand history. The account of Genesis is a historical narrative, not a myth, designed to situate us in history and fundamental issues of faith.

And when God says that he created the heavens and the Earth, this provides us with a starting point for both scientific and theological study. The scientific perspective that the Earth is approximately 4.5 billion years old is more solid than the idea that it is just over six thousand years old. It is a well-founded theory, and we will explain why. The principle that God created the heavens and the Earth does not give us a precise indication of the time involved in this process. Therefore, we should not dwell on this issue. If science presents new discoveries, such as a different age for the Earth, there is no theological problem with that based on Genesis. Dinosaurs lived millions of years ago, and this does not contradict the Scriptures. God could have created the Earth with an appearance of age, which also does not affect our theological understanding. We are facing a sovereign God, and we must stop limiting his power. Science is here to help us grow and explore the complexity of God's creation.

If you don't value science, then why do you use cars, medicines, laptops, and so many other things created by science? Science is here to develop technology and improve human health. We should encourage our children to study and become scientists, astronauts, or any other profession that motivates them to explore and discover. Unfortunately, many evangelicals, influenced by false doctrines, end up undervaluing science. This happens mainly due to theological ignorance, especially on the part of the Roman church's heritage from the past, which persecuted and killed scientists in the name of God. However, it is important for atheists to understand that it was not God who committed such acts, and the mistakes made by this church did not represent the thoughts of

Jesus and the apostles of the church, as evidenced in the New Testament. God cannot control everything that is done in his name, nor what is not. Everyone has the right to freedom of choice about their own decisions, and this is not related to election or predestination. It must be recognized that the representatives of this church could not perform acts or speak on behalf of the entire Christian community of that time, as no one can speak for a community or act on its behalf without its consent, which can lead to misunderstandings in history and distortions of values.

Therefore, we must be attentive to this. The name of God is important. I believe that the declaration of Genesis chapter 1, verse 1 outlines the nature of God, Elohim, the creator God, the purpose of God, and highlights his role as agent and creator of the cosmos. He decided to create and is self-existent. That question your child asks, God created you and who created God? He is self-existent, the one who reveals himself. He is our God, the one who exists by himself, that is why he is God. That is why he is our Lord.

He is the Creator and occupies himself as Creator. That is what the text of Genesis 1, chapter 1 says. He is the Creator and occupied himself with the activity of Creator, creating the heavens, the earth, the stars, the cosmos, and everything in them. He is the Creator, the self-existent one. He occupies himself in this role. Therefore, when Genesis chapter 1, verse 1 says that God created the heavens and the earth, it means that he occupied himself in the role of creator.

He worked as a creator. God had an occupation, so to speak. But the Scriptures say that he worked six days and on the seventh, he rested. He continues working to this day, God works. If God works, just imagine, that is why Prov 6:6 is right: whoever is lazy will be reprimanded. Our God occupies himself, he works, he is not lazy. He created and continues working in our lives. He occupies himself as a worker, creator. He creates with purpose because he creates for himself. The Scriptures we read earlier say that he created everything for himself, for his glory, honor, and praise. God creates for himself. You were created for God, I was created for God. We belong to God. The Earth, the heavens, the cosmos, he created for himself.

He appointed us as stewards, vice-regents, to enjoy and administer. Within his kingdom, he placed us on this Earth in his image and likeness, just like him, to manage, care for, and exercise dominion over this earth. But he created everything for himself. Do you think you have something? I think I have something? He creates everything for himself.

The Scriptures say that everything is his. Everything belongs to him. The devil belongs to God. The sinner who goes to hell belongs to God. He is even the owner of hell. It is a very broad declaration in Genesis chapter 1, verse 1. For it brings reference to time, in the beginning. In this context, the theory about the Earth comes in. In the beginning, God created the heavens and the earth. Here you enter a very broad context, which you cannot measure. Then comes the narrative. It is like an outline for the subsequent narrative. So when it says in the beginning, God created the heavens and the earth, you cannot measure. Leave that to science. This is not the work of theology. Theology will affirm that God created the heavens and the earth conclusively. If science studies atoms, particles, elements, compounds, metals, dust, it is important, it is scientific, but if our question is to understand that God created the heavens and the earth conclusively, there is no discussion. You don't need to argue with anyone, with the teacher, with anyone, because you already know that God created. This is really important for science to study. As I said, it's like a computer. One company creates a part, another creates another. The processor, the screen, the keyboard, another creates the software.

Investigating is part of science, it's inherent to humanity. So, when the Scriptures affirm that God created the heavens and the earth, if cosmology were concerned with confirming that this statement is true, it wouldn't be creating science, but should study applying the scientific methodology, without the purpose of proving that God created the universe. The Scriptures are also not concerned with proving the existence of God; if it were true, Moses would certainly have mentioned it, but there was no such concern. The problem lies in how people who don't believe in God begin to study the Scriptures. They are not concerned with the issue of their faith. The Scriptures say that by faith we understand, in Heb 11, that everything that exists was created by the word of God. This is a matter of faith, not a matter of theory, discussion, or physics. "In the beginning, God created the heavens and the earth." Now, if we are going to study all matter and particles, in that sense, it warrants a good discussion. Now, the question of whether God is the creator cannot be debated. You stick to your position, and whoever believes in the other theory sticks to their position. The big bang? There's no problem in believing that God, through what was speaking in his word, was causing explosions to bring about the universe, the whole cosmos. But the Scriptures leave no clues for or against it. To be honest, I believe it was like that. When God spoke, there was a big explosion. Any problem with that? If there is, show me in the Scriptures something against it. There isn't? So, like this, we waste a lot of time debating issues that are irrelevant to our faith. In certain situations, the enemy uses this to undermine the faith of many people.

Genesis 1:1 is comprehensive because it's talking about the beginning of time. The text opens by saying "in the beginning." Without this assertion, we would have serious problems interpreting the Old Testament text. It starts with a narrative about time. Now, why is that? This is a statement that takes us back before time, before the creative activity. If "in the beginning" is a narrative that points to time, it makes us think before time. Here is where God began creation. And one thing I reflect on a lot is the issue of time. Time, as it is, we understand it from the perspective of the fall. Adam and Eve, before sinning, understood in a different way than you do; they lived in eternity. They were eternal. They weren't bound by hours, minutes, seconds, and years. They were eternal, just as God is eternal. So, they weren't bound by this time. That's why there's a lot of confusion, a lot of conflict about this.

If we consider God's timeless nature, whose existence transcends the concept of time, we understand that the temporal notion we know is relative and limited. Within this perspective, the biblical narrative of the "beginning" in Gen 1:1 refers us to a reality that precedes time itself. In this context, it's important to note that the concern with time, as we experience it now, is a consequence of human fall, as reported in the fall narrative in the book of Genesis. Before the fall, Adam and Eve existed in a state of eternity, not bound by the temporal limitations we know today.

As we reflect on this, we realize that scientific investigation into the origin of the universe is a human attempt to understand the complexity of divine creation. However, this investigation is often motivated by seeking answers that exclude the possibility of God's existence. From this assumption, scientists strive to find explanations that do not consider divine intervention in the creation of the cosmos.

However, it's crucial to recognize that the narrative of Gen 1:1 transcends the limitations of science and points to a reality beyond human domain. The theological approach to this biblical text invites us to contemplate the greatness and glory of God as the Creator of the universe. Therefore, the interpretation of this verse should be enriched not only by scientific analysis but also by a deeper theological understanding that acknowledges the sovereignty and majesty of God over all creation.

There's always much to be explored, as the popular saying goes. The opening statement in Gen 1:1 establishes the starting point within the narrative of divine creation. When it mentions the "beginning," it refers to the inaugural moment of God's creative action, a point that transcends the very idea of time. In fact, any previous temporal conception must be understood within the context of eternity. As we study time, we realize that it's difficult to separate these two concepts. The concept of time as something in motion is particularly relevant due to our mortal nature.

In this context, the text speaks about the beginning of time in the creation account, when God created the heavens and the earth. Moses elaborated this narrative to express a synthesis of temporal beginnings. When exploring the Hebrew, we find a significant expression that God used when speaking to Moses in Exod 3:14. He said, "I am who I am." This translation, "I will be what I will be," reveals the eternal and self-existent nature of God, transcending the concept of linear time, pointing to the continuity of the divine Being beyond temporal limits.

The expression "I am who I am," and its expanded version "I will be what I will be," uttered by God to Moses, reveals the self-existent, eternal, and unchanging nature of the divine Being. In fact, this translation reflects the semantic richness of the Hebrew expression known as the tetragrammaton, YHWH. This verbal form suggests a temporal continuity that transcends the human concept of past, present, and future. God's choice of this phrase emphasizes his permanence, his faithfulness, and his ability to fulfill his promises over time. He is the Being who exists by himself and who is always present, regardless of circumstances and temporal changes.

There are many theories about the existence of God, some dating back to the times of Augustine. Nowadays, there is much discussion about this issue, and Hebrew writings portray God as the one who will come to be. One of these theories, originating since the times of the Greeks, possibly with Aristotle, proposes that God is the first cause, the prime mover from which all things arise. However, a postmodern theory questions this idea, arguing that God cannot be the first cause because that would imply an infinite series of successive causes. The Hebrew expression "I will be what I will be," spoken by God to Moses, addresses this issue, suggesting that God is beyond the concept of cause and effect. Therefore, the idea that God is the first cause is not correct because God transcends this notion and simply is what he is.

Many "Christian thinkers," in their rational attempt to convince unbelievers of the possible existence of God through logical arguments of reason, end up losing the logic of reason by using merely speculative

syntheses to prove God's existence. Is it reasonable to believe in the premises of metaphysics rationally developed about the existence of God? Can
our philosophy supported by rational metaphysics present a convincing
reason about the existence of God? After thousands of years of reflection,
do we have an irresistible argument in favor of his existence?

I am fully convinced that it is not possible to be a believer without
basing faith according to the Holy Scriptures, and merely forcing belief
in God from the vain thoughts of the human mind, which cannot fathom
the depths of God. If belief requires blind proofs from the science of
moving things, then faith is no longer vital to belief, but science is. And
if we believe in God without reasons and without logical facts, like the
irrefutable resurrection of Christ, we lose all proofs of our rationality and
common sense.

Many believers, well-intentioned but mistaken in their vanities,
think they can construct arguments to prove God's existence. Over the
years, many theories and theological currents have been elaborated for
this purpose, but they only created religious mazes. Among these theories, the Melchizedek factor stands out, namely, the philosophical argument that God must exist because, in almost all societies and cultures,
people share the belief in his existence. We can easily refute this argument in the light of truth.

A false belief can have universal acceptance; a large number of
people believing in a religion does not make it true, nor does it mean
that it presents an unquestionable theological foundation to prove something about God. Belief is not true simply because everyone believes it.
Truth must rely on pillars and evidence. The resurrection of Jesus Christ
seems to me a historical fact, although not universally believed, we have
historical evidence to accept it as true. We have unquestionable examples
in this existence that many false beliefs were or are almost universal,
without scientific questioning for a long time, such as the belief that the
Sun revolved around the Earth, or the scientific belief in the infinitude of
the universe, now seriously refuted. Although Islam is monotheistic, like
Judaism and Christianity, and is universally widespread, there are theological premises that approximate them, but also those that distinguish
them almost completely regarding the concept of God.

Although Judaism and Christianity merge on a certain theological
axis, their God is not what most people believe. On the other hand, there
is not the slightest possibility for a practitioner of Islam to justify their
belief in the Christian God. Every good Christian, with reasons for their

faith, is convinced, not by good philosophical and rational arguments about the existence of God, but by the greater proof that is the historical affirmation of the manifestation of Jesus Christ as the only begotten son of God incarnated among men. Their theology is entirely rooted in the irrefutable fact of the resurrection of Jesus Christ as eternal hope.

Many believers, doubtful of their own faith in the resurrection of Jesus Christ, find themselves in serious trouble of belief. Especially those weak in their metaphysical convictions, as they intertwined themselves in belief based on metaphysical arguments. Agnostics were a sect, if we can call them that, with their indecisive assertions that we cannot prove that God exists, or simply if he exists, he is certainly an unknowable being. If we cannot prove that God does not exist, then rational believers, without faith in the ineffability of the Scriptures, are also not authorized to believe that he exists from what was developed as the theory of God's existence. God's existence cannot be proven by theories at all; I find this absurd. The proof of God's existence by observations of science is not valid for faith in Christ, if I do not believe in him according to the premises of the Scriptures. Any logical argument in favor of God's existence that is not based on the word of God only strengthens the unbeliever, as they do not believe in what the Bible says about faith in God. The grounded atheist knows that faith is not rationalized; it is the strong point of the true Christian; therefore, we do not need to dedicate our convictions to the concept of a rational God to prove skeptics. Atheists claim that we cannot prove that God exists from what we consider biblical evidence, which is actually historical evidence for us, but they also cannot prove the non-existence of God with not even scientific arguments. They feel authorized to say that the Bible is a book of fables and doubt the faith that also proposes a relational experience with God, but they cannot explain with convincing evidence and with their evidence of disbelief the non-existence of God.

I am presenting a God who demands faith in the word, not inventing a belief in the universe. This is the best way to show that we are correct about the existence of God. The analyses of the cosmos from a skeptical point of view point to a possible emergence by chance, by itself, but even if this were true, skeptics have no proof against the existence of God, as he is also the master of chance. Therefore, let's once and for all annul the fallacy of chance. If something is the result of chance, it lacks a source of explanation, and trying to explain the inexplicable has no logical basis.

The most obvious thing would be to accept the mysteries of creation; the rest is just a futile exercise of reason.

Is it difficult to accept the deliberation of a universe brought into existence by some almighty deity? If the reader is someone who does not believe in God, they will have to reflect deeply on this issue, as it is almost impossible for the universe to have come into existence on its own, just as it is impossible for it to have existed for an infinite amount of time. At some point, everyone ends up slipping and believing for some reason, and then pride does not allow them to go back to continue believing in this possibility. We have strong reasons to frequently give credibility to God as the creator of the universe as we know it.

Over the years, countless theories and scientific arguments have reflected in theology, attempting to prove God's relationship with the universe, they are the cosmological proofs of God's existence. Within the logical argumentation of reason, they were outlined in the footsteps of philosophy to prove that there must be a "first cause" of the entire universe. If this first cause exists, we certainly do not prove the existence of God. He cannot be the first cause, he is the I Am, without beginning or end, as declared in the Holy Scriptures to Moses, and later manifested in the declarations of Jesus Christ in the Gospel of John.

We do not have the conditions to propose the ineffable description of this omnipotent being; we use human anthropomorphisms as the first cause. However, this conception of explaining cosmological things and events from cause and effect becomes irrelevant when we truly understand the message of the Tetragrammaton YHWH. It becomes almost impossible to continue trying to seek coherence of the existence of everything in terms of cause and effect, for we begin to believe in the God who follows the coming to be, becoming what is. He is the Cause to Come to Be, or to Show to Be, because that would mean that there should not be an infinite series of causes, which seems impossible. God simply is; he said to Moses: I am the One who is, and will be what I will be, the unexpressed, the self-existent, I am what shows to be. Because it is really difficult to understand this concept with our finite mind, some disapprove of the belief in the God of the Scriptures. But can anyone on this planet conceptualize eternity in their mind? Of course we cannot, but can we say that it exists? How can we prove if it exists? And if I do not believe in eternity, what do I understand by infinity? If there is a God that can be explained by the science of moving things, certainly that God does not exist. If we do not have full conditions to measure eternity, would we be foolish to think that it exists? Why do men think they can explain God in their limitations? We should simply believe by faith, just as we believe in the existence of eternity itself, even without full certainty of it, for it only makes sense because we exist, but eternity will not cease to exist because we do not accept it in it, just as God will not cease to exist because we do not believe in him.

Everyone, through knowledge, can perceive the order of the universe and imagine in the mind a designer as an architect, but with that, we cannot say that God exists. We can only affirm that God exists through faith in Jesus Christ because something that cannot be explained cannot be touched. These fermented theories always depend on the laws of physics and chemistry to be accepted, just like the proposal of the evolution of species, however, the Christian does not depend on a mathematical explanation to believe in God, for the eternal God cannot be proven by the speculation of the human mind. Therefore, he requires the path of faith in Jesus Christ to be known.

We do not have a formula to equate and describe the meaning of the designation I Am Who I Am. We are poor mortals, just as Moses was; reason cannot conceive the infinitude of I Am Who I Am. Can an enlightened person explain this to me? If they can explain it to me, then they are a god. Exodus 3:14: "And God said to Moses: I Am Who I Am. He also said: Thus you shall say to the children of Israel: I Am has sent me to you."

We do not need to try to explain its origins or attribute it as the first creating cause of all things, as proposed by the theory of the first mover, from where all things arose. God is what he is. He is the one who will come to be. God cannot be explained. In fact, he does not even have a specific name. With each manifestation of God, people gave him a corresponding name. When the people of God won a battle, He was called "Alvar-Nissim," our banner. If God brought healing, he was known as "Il-Alvar-Rafa," the Lord who heals. When he provided for hunger, he was called "El-Shadai," the one who nourishes us. In truth, God does not have a fixed name. The rabbis used the expression "Adonai", but the true name of God in all the Scriptures is the tetragrammaton, composed of four letters. There is no specific name for God. It is through his actions that the prophets and the fathers were giving him names.

I always affirm that the best way to refer to our God is as the Scriptures teach us: he is the God of Israel. He is Yahweh Rapha, he is Adonai, he is. He is the Eschaton in creation, a term derived from the Greek ἔσχατον, which not only denotes the end of the creative process, but also suggests that God began his work from the end, revealing his sovereignty over time and his eternal purpose from the beginning. It is very interesting because it is intrinsically linked to creation. It may seem a bit confusing at first glance, but this expression is in the Bible, so it needs to be taught, for if it is in the Bible, God spoke. There is a tradition that says the ancient Jews or scribes decided to hide the true name of God in the Scriptures, fearing that the Gentiles would use this name in vain and blaspheme against the name of God. That is why they put the tetragrammaton as an identification with I AM. But today, if you go to a synagogue, you will see that they are using the same names that we, Christians, use, which are in the Bible. There is no secret and hidden name. God reveals Himself as "I am who I am," the one who will come to be so. Before time, he showed himself as the one who would come to exist. He rules over what does not exist; God walked through nothing, nothing offered him praises, there were no particles, wind, rain, nothing.

God has always existed. He is self-existent, the Creator of heaven and earth, par excellence. It is shocking to think that he brings everything into existence from nothing, just with the word of his mouth. Reflecting on the attributes of God, something very important stands out: he allows us to conceive the attribute of eternity, which goes beyond time. Even though we are limited and finite in our minds, we can understand what eternity is. Why? Simply because it is something that God has allowed us to understand, even though it is something we do not possess. In relation to the other attributes, the Scriptures affirm that God is just.

In theology, it is recognized that human understanding of justice, truth, and love is often limited and influenced by various factors, including religious beliefs and the means that shaped the intellectual formation of human beings over time. Although God, in his perfection, is the apex of these attributes, our understanding of these concepts is often obscured by our own limitations and by the external influences that shape our perceptions.

The majority of spiritual beliefs and traditions revived in current practices have resurged from mass misunderstandings, and without hesitation, are accepted by many of us without question. Almost all are uninspired sources of some sort of naïve inheritance from our ancestors who lacked a critical spirit. These cultural rites were almost all forged by the natural yearning for a connection with the knowledge of God. This is often spiritualized into rituals without thorough theological scrutiny.

The quest for spirituality often leads to the resurgence of ancient practices, some of which included deeply disturbing rituals. It is evident in various cultures that throughout history there has been a rebirth of the scattered fragments of ancient polytheistic beliefs. These beliefs, which in some cases date back to times when human sacrifices, including those of children, were accepted as offerings to appease the gods or influence the spiritual world, have left traces that are sometimes revisited in modern times.

In pre-Columbian Mesoamerica, for example, civilizations such as the Aztec and Maya are notorious for their practices of human sacrifices. Such rituals were seen as essential to maintain cosmic balance and the favor of the gods. At the height of the Aztec Empire, ceremonies involving sacrifices were elaborate and fundamental to various religious festivities. Today, some of these beliefs and rituals survive in hybrid cultural and religious practices, although human sacrifices have been abandoned, replaced by symbolism and metaphorical rituals.

Similarly, in ancient Carthage and in some Canaanite cultures, child sacrifice was practiced in honor of deities such as Baal and Moloch. Archaeological excavations have revealed burial sites where the remains of burned children were found, suggesting sacrifice rituals. While these practices have been extinguished, reflections on such acts serve as a dark reminder of extreme expressions of devotion in human history.

In Europe, during the pre-Christian era, there are also reports of human sacrifices practiced by various pagan tribes, such as the Celts. Druids, the Celtic priests, supposedly performed human sacrifices in complex ceremonies, often associated with harvest festivals and transitional moments. While modern Celtic practice, or Neopaganism, vehemently renounces such acts, seeking to reconnect with nature and the seasons, its origins date back to a time when the value of human life could be subjugated to the perceived demands of the divine.

These practices, now viewed as abominable, highlight the evolution of human understanding of the sacred. The contemporary view of ancient religions often selects positive and philosophical aspects, leaving behind the more brutal and unacceptable components, transforming them into allegories or dismissing them as a misguided understanding of the divine.

I am convinced that our problem lies in our foolishness of having various opinions about the same object of focus only to prove our vanity rather than the truth, and also to follow blind faith in doctrines without verifying them. The conspiracy of some senseless agnostic atheists is the result of the frustrations of these inheritances and mental prognoses partly used to frighten those discredited in the possible promise of the existence of heaven.

Using our religious customs is a deadly and corrosive prison for our minds. I say this with the caution of reflection without any kind of pity or excuse, for in the name of belief in God, men have added deaths, incalculable violence, and insane customs to lead the ignorant flocks of truth to blind obedience.

Jesus Christ issued serious warnings to ancient sects, including the Pharisees, against this disease of using political power for the purpose of controlling the poor defenseless flock of lean sheep of the Judaism practiced at that time. "Why do your disciples break the tradition of the elders? For they do not wash their hands when they eat bread." He answered and said to them, "Why do you also transgress the commandment of God because of your tradition?"

"For God commanded, saying, 'Honor your father and your mother'; and, 'He who curses father or mother, let him be put to death.' But you say, 'Whoever says to his father or mother, "Whatever profit you might have received from me is a gift to God"—then he need not honor his father or mother.' Thus you have made the commandment of God of no effect by your tradition. Hypocrites! Well did Isaiah prophesy about you, saying: 'These people draw near to Me with their mouth, and honor Me with their lips, but their heart is far from Me. And in vain they worship Me, teaching as doctrines the commandments of men (Matt 15:1–9).'"

Although the beliefs of Judaism were deeply affected by the religious attitude of this group, Jesus made it clear that the word of God remains, although that culture was customizing a lifestyle in people. They were being influenced by the environment created by the Pharisees as a false proposal of God's ordinances. It is very dangerous when individuals live their faith under strong influences of customs and cultures, as this can stifle the truth, for we are almost always intellectually and spiritually shaped by the environment around us.

But it is not through this path of going with debris that we will return to a denial, saying it is not right in detriment to the deceit spread on the ground by obstinate men. We are aware, without questioning much, of how an environment can produce spirits intertwined with its habits and customs, but to affirm that being born in a certain place will radically influence our decision with God is to be ignorant of the theological and philosophical concept of God's manifestation. One thing is the presentation of a god as a belief proposition to a people where he is unknown, and another, totally different, is his incarnation in that people as the living God.

The religion of parents is a myth and often extends into the lives of their children, but it is not true that all children are manipulated by their parents' religion, and this is known to the majority of us, so I do not need to exert myself to defend this idea, thus wasting my time on reflections that require greater use of reason. The Irish are not Catholics because their parents are Catholics, just as the Calvinists of Scotland did not choose this faith because of their ancestors' religion. So, do we assert that man is religious because there is religion? Or is there religion because men are religious? A good question would also be whether man needs religion. If men are religious because they are religious, would it then be wrong for them to seek a religious practice? A good conclusion is that religion exists because of man, and not man because of religion. For some reason, we can affirm that he is religious due to the *sensus divinus*, but believing in God is an intrinsically human necessity, as the flames of atheism arise from the absence of this satisfaction found by many in their spirit.

Being an atheist is defined as someone who does not want to believe in God, and in order not to be an atheist, he needs negative faith in God. However, to do this, he needs to deny the very essence of his negative thought that God does not exist. It is easier for a camel to pass through the eye of a needle than for an atheist to convert to the existence of God, as this will undermine their unscrupulous conduct, and that is what they truly cling to. In doing so, they create a general rule, with few exceptions and divergences of ideas, and without realizing it, they are all blind to think that they are not also following each other to build defense mechanisms against those who believe in God. If they call religious believers insane, they should also investigate the serious ideological insanity of a world that they have never been able to build, and are living in this exaggerated non-religious faith as if they were in the land of dreams and do not want to wake up to the human reality of hardships.

The intellectual formation of an atheist can begin by being justified in the act of religious imposition by their parents, although it is not particularly wrong to teach children the way to go. Saying otherwise about this issue would also be the same as going against certain traditions and cultures that are passed down from generation to generation. On many of these issues, we do not perceive severe attacks from opponents of belief in

God, if in fact the majority of these neo-atheists were stigmatized at some point in their lives by the religious impositions of someone.

Tyranny, in analysis, is not much different from rape or any other type of violence imposed by force. Now, teaching children does not make them victims of any kind of ingrained belief, just as we cannot say that this happens in the choice of a soccer team. We are correct about many things acquired through experience over the years, but we are not entirely certain about many generalized assertions regarding the transfer of religious values as a prison without keys to freedom of choice. We are not all culturally religious, although some severe religions have mental cauterization as a rule, making them insane. These people obstinate in blind belief, without the right to question their faith, differ in various aspects from the human rights of Christianized peoples, where freedom is not assisted. Atheists should thank Jesus Christ and Christianity for the free life they lead, instead of remaining on the sidelines of only negative investigations.

I am not afraid to err in my propositions; if it were not for Christianity, the world would not be as it is, and would never become what it will be, whether for the good or the bad side of history. Where it predominated with force, the greatest scientific and cultural development occurred, along with freedom of expression; this is a historical fact worthy of evaluation. If there is a religion that delayed scientific-cultural progress and knowledge in general, it was not Christianity, although some skeptics evolve their criticisms due to the persecutions that occurred in the Dark Ages, which is a dark period of Christianity, a victim of ignorance.

The errors of the church are the errors of men, just as many philosophical thoughts were wrong. It is up to us to judge and clothe the truth that was laid bare by the lies embraced as dogmas in the name of Christianity, but let us understand that this was the only form of power known to the clergy. This bizarre source of illusion collapsed in the light of truth and made Christianity stronger; thus, it is not an emerging culture of human thought, but a living spirit of freedom, which can always start with the initiative of one, also of a group, but this does not make us products of the environment, as we can make decisions regarding belief or disbelief in God. Atheism itself corroborates this assertion, just as God is not a product of civilization, and our belief in him does not stem from the geographical and cultural position where we were born.

Saying that human thought is limited to the environment of interaction is the most ignorant demonstration regarding the history of humanity itself, from Socrates' justification in harmonious relation to the philosophical concept of "Zeus" in opposition to Greek mythology, to the theological contemplation of Ormuz by Zoroaster.

Religions for the most part are creations of the human imagination; truth is a differentiated attestation, although there are truths in religious philosophies of the imagination as in many scientific creations. Only God is the truth, and it does not incorporate itself into our modeling, for truth is free, it is Jesus incarnating as the living word and resurrected as God of the universe.

Belief springs forth like a clean source gushing down mountains and can become dirty in the course of receiving strange doctrines used as sculptors of a new way of thinking for the individual. This is independent, for if he followed another pattern of thought, he will need discernment to ensure the truth as the rule of faith as the proposed message of hope of Christianity. This will certainly subject itself to the obstinate wills of the human ego and lead it to the spiritual model for a life free from the oppression of the spirit. Not always are men endowed with the pure faculties of the spirit, conceiving more refined thoughts regarding truth as the objective of belief, hence the need to teach the truth.

Some believe, for example, that children do not understand as adults do when receiving the same message. This fallacy would be equivalent to saying that Spider-Man, Superman, and Batman became different superheroes in my current conception because I grew up, but I see them in the same way. The view I have of them regarding their heroic acts and almost absolute powers still has the same meaning in my rational imagination. I see them little different from what I believed about them when I was a child. In fact, we all see them like this, we are intertwined with the world of superheroes, although even as pure honest children, we knew that it was unreal. Children can believe as we adults do, and when they mature, they can better refine the path of their belief, but believing represents the same sense as their parents, since they believe exactly what they were taught. Although it is correct to say that their worldview is not entirely identical to that of their parents, this does not interfere much with the standpoint of belief, only when there is some kind of psychological manipulation by the parents.

Neo-atheists feel uncomfortable with the religious education of children because most of them were victims of some kind of religious abuse and did not have the opportunity to decide what to believe in. With this, they want to deny children the right to know the love of God. How many of them, professionally, became what was planned by their parents and are now frustrated? How many took unwanted courses and studied in schools due to their parents' decisions? Many wore clothes chosen by adults, and even ate foods they didn't like, simply in the name of good health.

The universe of a child is more extensive than we imagine. Setting specific guidelines like the good path of God may seem somewhat illegitimate, but we do many other things with children in the name of correctness that also do not seem legitimate to me. The most important thing is to see the world we are building within them.

The atheist says that the religious world is an imposition that fosters prejudice, but what is the atheist's attitude towards the believer in God? Typically, in their "remarkable findings," we perceive expressions of contempt, pretension of intellectual superiority, and snobbery that believers are ignorant, and the most vulgar and prejudiced classify believers in God as insane fools. The psychotic respect preached by them regarding the prerogative of the end of religion is also a tyrannical act of love for ideas, instead of directing it towards people.

For many atheists, banning faith is a demonstration of holding absolute truth that they themselves deny, thus making denial an absolute act. If they do not have absolute truth, why bring to the world another ideology rooted only in the expression of hatred and intellectual vanity? Atheists like Richard Dawkins put scientific development as the precursor messiah of the new non-religious order without acknowledging that science ironically arose from religion. So, it is not very intelligent to treat science as if it were an enemy of religion. It is strange to talk about conscious and unconscious growth; and to unearth comparisons to ferment offensive dissensions so that people believe without hesitation that the old religious books should be completely abandoned or perhaps entirely supplanted by the new scientific world, instead of trying to promote peace between the two branches.

Fortunately, man is capable of change, and within a few years his mediocre thoughts may be different. The human mind hates being stagnant and forgotten, so we desire progress, but men like Richard Dawkins have made us regress to slavery to lies in the name of truth. Therefore, neo-atheism is dwindling in its own organism without a proposal for life. The latest virtual offenses are the most vulgar form of escape from this almost certain death, as if humiliating people who believe in God with rude expressions made them stronger. This is not the best way to be the promoters of a better world. The impressions of creation cannot be the defining mark of our religious character, knowing that the influences absorbed in social settings can also shape a neurotic character in children, beginning with negative model contacts experienced by the religion of their parents. We cannot live religion as if it were the truth, if deep down we know it to be a lie, doing so only to fulfill an agenda of commitment to avoid saddening others. God is not granted solely through education about him, as a particular spiritual experience of each individual is necessary. If the focus of belief is only on the results presented by the environment, there will certainly be great frustration, even if we receive a good religious education, it is necessary for people to discover the true reason for the love of God, as many of them have lived the conflict of misunderstanding derived from the opposition between God's justice and love. As a result, they mainly read in the Old Testament (which is the sacred book that reports the divine plan for the education of Hebrew slaves) texts and could not clearly understand. In these passages of the Old Testament, Jehovah is as measured in compassion as he is extravagant in the aspect of violence against nations imagined by us as victims. Thus, in the mind devoid of theological knowledge, God becomes a being eager to kill, astute and cold in the diligent mode of murdering his prey without mercy. The end product of this artificial reality in the conception of doubtful people of God's goodness is like an arrow that will lead them to mortal hatred in their hearts against God. How will these doubtful individuals be able to see the desolations, the catastrophes where innocent babies were killed?

On the pages of the Bible, many women can be seen being violated in their feminine rights, and old men who had their clothes stained with blood. We cannot deny that perhaps the Bible is the book where blood flows the most through its pages, from the murder of Cain to the death of Jesus Christ and his followers. My faithful assertion is: the Bible is the most faithful book that exists, it did not hide the acts of God, but rather printed them on paper. It is the book that also did not hide the evil of human beings in an attempt to convince people of a falsified human being under the pretext of goodness. Certainly, the Bible is the black mirror of man and the thorns of the branches of the rose called God. It is the book most faithful to its characters written in the history of humanity. If the Bible were invented to convince people only of the love and kindness of God, would it be written with verses like this? The way it is written is really to cultivate faith and bring understanding of a just God. Some atheists discredit the love of Jesus Christ because of the justice and condemnation of hell. If these men and women of disbelief were in Hitler's concentration camps, would they desire some kind of justice? At least if their sons and daughters had been crushed on September 11th by Bin Laden, what would their sense of justice be? They find hellfire heavy for terrorists, so they got away with this joke of judgment, as there is no judgment. What do atheists say about women who have been abused and become stigmatized, and also boys and girls who have been sexually violated? From what we know, God does not commit sexual violence, but men and women do it through the usurpation of religious powers without God.

The Bible is the book of hope for the unjust, so our obvious conclusion will be the existence of hell. If there is no standard of judgment and divine imprisonment, in whose name are we imprisoning criminals? In the name of justice? What is justice? Did it arise from the moral refinement acquired through community living? So is it fair to condemn me for living my natural state as an animal? If morality is a state of evolution and comes with it, then I am condemned for being an animal that has not yet developed morally? If morality is taught through education, in whose name do people, endowed with that privilege, commit their barbaric atrocities? Animals devour each other and there is no consciousness of innocence and guilt, it is a terrible mistake in the evolution of the species. Religious and atheists are committing crimes and mothers are losing their murdered children, yet we are sleeping on the sidelines of the evolution of the species. What kind of stupid species is man attributed to? The Bible is so real in presenting this perverse animal that it created severe codes of law and repression to the point of killing them, yet it did not work to establish a community without evils and violence. Now our only hope is the fire of hell, but I see that men no longer care about it. If Jesus Christ, hellfire, and the Bible are a book of treacherous lies and fables full of raw violence in the name of God, but carried out by rebellious humans, in whose name will there be fear and reverence in this disorderly and evil world? "If God is dead, everything is allowed." If so, no one should condemn anyone for being an animal falling back into essence, because according to atheists, the laws of the Old Testament are severe and poisonous and such a God does not exist, and if he exists he should not be worshiped, but they forgot that they were laws created for men slaves of a "species" in the process of "moral evolution."

If we follow the premise of many atheists, all the evils practiced in the past follow the same process of collective moral improvement, so we must consider all the diabolical men in the history of humanity as innocent in moral evolution. True justice says: God exists, man is a moral being in essence, and must be judged for all his immoral acts, with the penalty of being responsible for them. And if he manages to escape the justice of men, in which we defend and honor as good faith, he will certainly not escape divine justice. The punishment for such a penalty may be the fair judgment of hellfire as punishment for this possible reality. "In God we trust, because in him we live, and move, and exist; but God, not considering the times of ignorance, now announces to all men, and everywhere, that they repent; for he has determined a day when he will judge the world with justice, by the man he appointed; and of this he has given assurance to all, by raising him from the dead" (Acts 17:28, 31).

God, in his wisdom and generosity, has endowed us with divine attributes such as love, justice, and mercy, which we can manifest in our interactions with others. However, there is something he did not grant us: the ability to be eternal by our own will. This understanding is intrinsic to the divine nature and surpasses our finite condition.

While we can seek to love and be just in our lives, the idea of living eternally transcends our human understanding. This realization leads us to reflect on the true nature of existence and connects us to the need for a greater source of eternal life, such as that offered through Jesus Christ.

The question of eternity is something that resonates deeply within us, as it was planted in our hearts by God. We are made to yearn for eternity, and that is why the concept of eternal life is so understandable to us. By preaching the gospel and speaking of eternal life in Christ, we can find a universal resonance, as it is something that everyone can understand at our deepest essence.

On the other hand, the love of God is often misunderstood and questioned. People may doubt the divine love and justice in the face of the difficulties and sufferings they face in this world. However, the truth of eternal life is more easily accepted, even if some reject it, as it is a reality that resonates in our deepest consciousness.

By receiving Christ into your life, you are invited to partake in the eternal life that God has promised us. This is something that everyone can understand and appreciate, as it reflects our own nature created for eternity. It is a glorious promise that frees us from the concerns and burdens of this passing world.

Regarding creation, the Scriptures reveal that God was active from the beginning of time and will continue perfecting his work until its completion. Even though for God everything is complete from the outset, he still works to shape his entire creation perfectly. The term "Elohim" used in Genesis 1 suggests a creator who is active in multiple works, which is fascinating for those interested in exploring Hebrew themes in depth.

The term "Elohim" is truly intriguing, as it can be used as a title attributed to prophets and kings, indicating authority and representativeness. Furthermore, "Elohim" also suggests the idea of divinity and creative agency in multiple works. This leads us to a deeper understanding of the nature of God and his operation in the world.

When it comes to Jesus, the Messiah, he is referred to as "Elohim," highlighting his divinity, authority, and representativeness as the incarnate God. This may be a source of confusion for some, especially for Jewish rabbis, who emphasize the unity of God. However, for Christians, understanding Jesus as "Elohim" reveals his divine nature and his essential connection to the Father.

Indeed, the issue of the plural "Elohim" can generate confusion and interpretative challenges for Judaism. The phrase "let us make man in our image, according to our likeness" may seem to suggest the involvement of multiple entities in creation, raising questions about the nature of God and his relationship to other creating entities. This ambiguity may represent a theological dilemma for orthodox Jews, who emphasize the absolute unity of God and do not accept Yeshua as the pre-existing Son of God.

For Christians, however, this passage is interpreted in light of the Trinity, where the Father, the Son, and the Holy Spirit are involved in the creative work. This reflects a more complex understanding of the nature of God, which transcends the understanding of orthodox Judaism unbelieving in Christ as the Messiah.

He will not explain why "let us make man in our image, according to our likeness." This ambiguity is often addressed in sermons, where it is frequently argued that the Spirit of God and God are one person. So why the term "let us"? And why the use of "Elohim"? Jesus, as described in the Gospel of John, is considered the Word, the Logos, present from the beginning of creation. For us, this does not represent a contradiction.

For we possess the complete revelation of God. The Old Testament alone is not enough; without the New Testament, understanding is limited. Without this complete revelation, questions such as the nature of the Messiah, the agent of creation, become confusing. If we do not understand that the Messiah is part of the Trinity—Father, Son, and Holy Spirit—we risk misinterpretation.

If angels participated in creation, we must question traditional theology. But our understanding comes from God's revelation. From the beginning, he is presented as the Creator of the heavens and the earth. This unity of the Trinity is fundamental. Furthermore, the eternal God, the Creator of all the Earth, does not grow weary or exhausted; his wisdom is unfathomable. Amen.

# Chapter II

## The Dead Universe Theory

IN THIS CHAPTER, WE explore a new frontier of cosmology: the theory of the dead universe. This theory proposes a revolutionary view of the origin and expansion of our cosmos from the perspective of the influence of the dead universe, suggesting that the cosmos in which we reside is influenced by the remnants of a previous universe. Unlike the conventional model that begins with the big bang explosion, our journey starts with the echo of a cosmos that no longer exists, whose fundamental laws and residual structures still shape the expansion or concentration of mass in the current universe.

The "dead universe" theory postulates the preexistence of a primordial cosmos, perhaps billions of times vaster than our known universe, which may have perished in the sense of stellar death, or perhaps, never departed from its innate state of nonexistence. In this construct, the proliferation of primordial black holes is not configured as a deviation, but as a constant inherent to the essence of this primordial universe. In contrast to theories that proclaim the origin of our universe as the result of collisions between ancestral black holes, this proposition advances the idea that the universe unfolding before our eyes is nothing more than a remnant, thin particles remaining from this defunct universe.

According to sacred scriptures, there is a "belief" that, at the origins of the universe, light was not initially present; it was subsequently created. This is corroborated by both the creationist view, which suggests the universe was shrouded in darkness until God declared "let there be light," and the scientific perspective of these primal events. Moreover, Gen 1:1–2 proposes the existence of a "previous universe" enveloped by a "black hole," surrounded by darkness prior to the existence of the observable universe. Therefore, it can be asserted that our universe resides within a black hole of this "dead universe," suggesting a complex interaction of spaces and dimensions before the cosmos as we know it was formed.

Therefore, this work may be scrutinized from various angles, but its essence, precision, and ultimate outcome will remain the same in its entirety. There is a singular destination for this work: humanity.

This cosmological approach offers a more comprehensive explanation of the origins of the universe, challenging conventional models such as the big bang, and all other theories associated with the expansion of the universe. In fact, this theory suggests a simple distancing of galaxies, as opposed to an aggressive expansion that would lead to the disintegration of matter and eventually the death of the universe, as proposed by theories of cosmic expansion.

Two hypotheses are proposed within the scope of the "dead universe" theory. Initially, the term "dead" is redefined, transcending the traditional notion of stellar extinction, to denote a universe whose fundamental characteristic from its conception is the intrinsic absence of light. In this model, light is considered a cosmic anomaly that arises from fusion and collision events between supermassive bodies within a primarily dark universe. Furthermore, this theory asserts that black holes and mergers are not the creators of the universe in which we reside.

The first hypothesis proposes that phenomena such as supermassive black holes, dark energy, and dark matter constitute the elementary components of this primordial universe. Interestingly, light appears under specific circumstances, possibly as a byproduct of complex gravitational interactions, acting as a catalyst for the transition to an illuminated cosmos similar to what we observe today.

The second hypothesis proposes that an ancestral universe, vastly larger than the currently known cosmos, serves as the final reliquary for the death that befell all galaxies and extinguished the light of a once vibrant universe. This predecessor universe could provide crucial evidence of the cosmological processes that culminated in the observable state of the universe.

The beauty of this theory lies in its ability to offer verifiable predictions. For instance, it suggests that certain anomalies in the universal expansion could be explained by the influence of the prior universe, presenting a new field of study for astronomical observations. Detailed analyses of the cosmic microwave background radiation or the distribution of dark matter could reveal unexpected patterns, serving as empirical evidence for the dead universe theory.

It is postulated that the observable universe already existed in a state of death trillions of years before the big bang. The emergence of light, triggered by intense activities in this dead universe, would be analogous to the peculiarity of black holes in the context of known physics. The theory of the "dead universe" implies that the cosmos we know is the residual aftermath of a vastness long gone, where the concept of stellar birth is reversed to universal death. In this scenario, black holes are not the catalysts of creation, but rather the epitaph of a universe that has exhausted its vitality. Instead of being generative singularities, these primordial black holes are the last gravitational beacons of a cosmos that no longer exists. The galaxies and stars we observe, in their apparent youth, are in fact the embers of a long-extinguished cosmic fire. Dark matter and dark energy, the enigmatic elements of our universe, can be interpreted as the faint echo of this final cataclysmic event.

Thus, galaxies and the universe existed, thus, in the absence of light, submerged in the darkness of the dead universe, but preserved in a state organized by the laws of physics. This perspective suggests that the universe did not merely emerge but has always been present in a state of death, previously in full darkness. This conjecture finds parallels with the "cosmic dark ages" proposed by the big bang theory, describing a prolonged period after the big bang but before the formation of the first stars, when the universe was pervaded by darkness.

Anomaly of Light: Light, a fundamental manifestation of electromagnetic energy, occupies a pivotal role in the physics of the universe as we know it. To propose that light is an anomaly in this theory is not simply to invoke complexity; rather, it offers answers to some of the most profound questions in classical physics. This approach does not just reinterpret established physical concepts but also proposes a new way to understand the nature of the universe.

The creation of light in stars is a complex process that primarily occurs through thermonuclear reactions in their cores.

Nuclear fusion: The primary mechanism for creating light in stars is nuclear fusion. In the stellar core, especially in stars like the Sun, hydrogen atoms are fused to form helium in a process called nuclear fusion.

During this fusion, a small fraction of the atoms' mass is converted into energy according to the famous equation of Einstein, $E=mc^2$. This energy is released in the form of light and heat.

Fusion cycle: In the Sun and other stars of similar size, the main fusion cycle is the proton-proton process, where four hydrogen nuclei combine to form a helium nucleus, releasing photons (particles of light) in the process.

Gravitational pressure: Nuclear fusion only occurs in stars due to the immense gravitational pressure in their cores, which forces the hydrogen nuclei to come close enough to overcome the electrical repulsion between them and allow fusion.

Hydrostatic equilibrium: The light generated by nuclear fusion exerts an outward pressure, balancing the force of gravity that is trying to compress the star. This hydrostatic equilibrium keeps the star stable and in its current state.

The scientific premise is that in the origins of the universe, light was not initially present; it was created subsequently. This is supported by both creationist beliefs, which suggest the universe was shrouded in darkness until God commanded "let there be light," and the scientific perspective that recognizes darkness preceded light in these primordial events.

Primitive Elements: While black holes, dark matter, and dark energy are well-established concepts in modern cosmology, they are often considered as emergent phenomena and not necessarily as primordial components of the universe. However, the theory of the dead universe offers a plausible explanation for their origins, presenting them as fundamental elements of a previously inert cosmos. Although dark matter and dark energy are areas of intense research and debate, with their origins still undefined by consensus, this theory presents one of the first rational approaches attempting to elucidate these enigmatic phenomena.

Expansion of Cosmic Understanding: These ideas challenge our imagination regarding the universe and provide fertile ground for theoretical discussions and speculative narratives. While remaining distant from the current scientific consensus, these theoretical considerations seek to expand our understanding of the possible states of the universe and the fundamental forces that govern its evolution and potential purpose. Thus, respecting the limitations of endorsed scientific knowledge, these propositions allow speculative exploration based on alternative theories and hypotheses.

The theory of the "dead universe" implies that the cosmos we know is the residual aftermath of a vastness long gone, where the concept of stellar birth is reversed to universal death. In this scenario, black holes are not the catalysts of creation, but rather the epitaph of a universe that has exhausted its vitality. Instead of being generative singularities, these primordial black holes are the last gravitational beacons of a cosmos that no longer exists. The galaxies and stars we observe, in their apparent youth, are in fact the embers of a long-extinguished cosmic fire. Dark matter and dark energy, the enigmatic elements of our universe, can be interpreted as the faint echo of this final cataclysmic event.

Among the theories describing the ultimate fate of the universe, hypotheses of the "Big Freeze," "Big Rip," "Big Crunch," and "Big Slurp" suggest dramatic scenarios based on the continuous expansion, contraction, or phase transitions of space-time. However, the theory of the "dead universe" presents a more serene and fundamentally different outcome for the cosmos.

According to the theory of the "dead universe," there is no cataclysmic or explosive event marking the end of the cosmos. Instead, the universe simply returns to its natural state, a state without the anomalies that characterize the observable universe. In this theory, light is considered an anomaly, something that does not spontaneously arise but requires nuclear fusions and other energetic processes to manifest. In this context, the existence of light is seen as a temporary disturbance in the primitive and eternally dark state of the universe.

From this perspective, if there were living beings in this primordial universe, they would consider light as something strange, a curiosity, or an intruder in the perennial state of darkness. The theory posits that, unlike stellar death, which can be a spectacular explosion, the universe does not end with a bang, but with a silent and inexorable return to darkness.

From a scientific point of view, life as we know it may have emerged from this anomaly, from a series of accidents and powerful mergers in the primitive universe that would ultimately draw the cosmos back to its native reality. The theory suggests that, just as light and life emerged from extraordinary events, the universe will one day reabsorb everything back into its original state, devoid of light and life, where dominant gravitational forces would facilitate this reversion to the primitive and dark state.

The theory of the "dead universe" challenges the conventional view of continuous expansion proposed by the big bang and other cosmological theories. It does not predict a violent or cataclysmic end, but a gradual decline into a silent equilibrium, where the cosmos slowly fades into the dark background of its primitive existence, remaining forever in a cycle of transient light and eventual darkness. The expansion of the universe, a phenomenon observed and described by Hubble's Law, shows that galaxies are moving away from each other at a speed proportional to their distance. This central fact of modern cosmology is in harmony with the theory of the "dead universe," albeit with a substantially different interpretation of the implications of this expansion.

In the theory of the "dead universe," the observed expansion is not the result of an initial impulse from an explosion, as in the big bang, but is seen as a simple distancing of galaxies due to the influence of gravity and other yet-to-be-understood laws emanating from the nature of the "dead universe" itself. This movement is interpreted as a manifestation of the intrinsic and residual properties of a cosmos that is no longer active in the traditional sense.

In other words, while Hubble's Law describes what we observe, the theory of the "dead universe" attempts to explain why we observe it. It suggests that the unknown laws of the "dead universe" may be residual forces or echoes of a previous cosmic reality, which now direct the dynamics of the observable universe. These forces could be different from the known classical gravity and could explain why galaxies continue to move apart even when the original energy of the big bang should have dissipated.

Therefore, the expansion would not be a sign of continuous growth or birth, but a gradual return to the quiescent and fundamental state of the "dead universe," a final state of rest after the end of anomalies like light and the complex structures that characterize our current universe. Thus, the theory of the "dead universe" adds a new layer of understanding to the ultimate fate of the cosmos and offers an intriguing counterpoint to prevailing cosmological theories.

Your description suggests an application of Newton's law of universal gravitation, which states that bodies with mass exert a gravitational attraction on each other, and this force is directly proportional to the masses of the bodies and inversely proportional to the square of the distance between their centers. According to the theory of the "dead universe" you are exploring, gravity would play a fundamental role in influencing the movement of the observable universe.

Adopting this perspective, we could theorize that a "dead universe" containing bodies of incalculable mass is exerting a gravitational attraction on our observable universe, pulling it back to a state of greater uniformity and tranquility. This would imply that the forces responsible for the expansion of the universe, such as dark energy, could eventually be countered or even overcome by the gravity of such cosmic superstructures.

To substantiate this theory, it would be appropriate to refer to various recognized physical principles and discoveries:

Newton's Law of Universal Gravitation: As a basis for understanding gravity on a large scale.

Einstein's General Relativity: This theory updated the understanding of gravity as the curvature of space-time. The role of singularities and event horizons in black holes could be explored in relation to the "dead universe."

Hubble's Laws and Observations from the Hubble Space Telescope: They provide empirical evidence of the expansion of the universe, which could be interpreted in light of the attraction of a massive "dead universe."

Quantum Cosmology: Investigating the implications of quantum mechanics on cosmological scales, this area could provide insights into how a "dead universe" could influence matter and energy in our universe.

Research on Dark Energy and Dark Matter: Studies on these mysterious components of the universe could be useful in understanding the forces that are competing with or interacting with gravity in the "dead universe."

Notable scientists for reference could include:

Stephen Hawking: For his work on singularities and the properties of black holes.

Roger Penrose: Who collaborated with Hawking and developed theories on the nature of space-time.

Saul Perlmutter, Brian P. Schmidt, and Adam G. Riess: Astronomers who were awarded the Nobel Prize for their discoveries regarding the acceleration of the expansion of the universe.

Kip Thorne: A theoretical physicist who made significant contributions to the understanding of gravitational waves and the nature of gravity.

Finally, it would be vital to conduct computational simulations and mathematical analyses to test the viability of this hypothesis, perhaps using modifications of Einstein's field equations to include terms representing the influences of a "dead universe." This would help provide a robust mathematical model that could be compared with astronomical observations to validate or refute this alternative theory about the fate of the universe.

Similar to stellar birth originating from dense cosmic clouds, our universe may have been partially forged from the remnants left by its predecessor. The younger stars and galaxies we observe, shining in their billions of years of existence, may very well be the ultimate creations of this defunct universe. On the brink of its extinction, it was still capable of originating new structures, suggesting that its cosmic end and the beginning of another process leading to final annihilation are completely interconnected, culminating in approximately hundreds of billions of years, observable facts in our current universe, governed by the immutable laws of conservation and metamorphosis that dominate all of natural reality.

These young galaxies can be understood as final echoes or shimmering memories of a cosmos that no longer exists. They are fragments of a vast stellar legacy, the last whisper of a universe that was once vast in scope and energetic wealth. Thus, we are living in the shadow of a glorious cosmic past, witnessing what can be considered the "last dance" of light and matter originating from a fading universe. What we perceive as our stellar reality is nothing more than a remnant—a small yet vibrant part of a much larger existence that extends beyond our temporal and spatial reach. In essence, everything that is, everything we see, and everything we may come to know are just preserved fragments in time and space, the indelible signature of the dead universe.

As this process unfolds, the density and complexity of the universe decline. Where there were once dense clusters of matter and energy, now remain increasingly vast and empty spaces, punctuated by isolated islands of stellar activity. The observation of young galaxies by the James Webb Space Telescope, then, serves as a glimpse into this declining process, revealing the final stages of the cosmos that we are only beginning to understand.

In this framework, the death of the ancestral universe was not an abrupt event but a prolonged phenomenon that allowed the gradual emergence of new structures from its ruin. Black holes, instead of being the catalysts for a new birth, are the final guardians of the cosmic memory of the predecessor universe, storing in their gravitational abysses the history of everything that ever existed.

Black holes indeed have mass. The mass of a black hole can be comparable to that of the Earth, the Sun, or even much larger, depending on the type of black hole. There are stellar black holes, which usually have masses of a few to tens of times the mass of the Sun, and supermassive black holes, which can have masses equivalent to millions or billions of times the mass of the Sun. The term "black hole" refers to the fact that these objects are regions of space where gravity is so intense that nothing, not even light, can escape from them. The word "hole" is a way to describe this "capture" characteristic, although it is not a hole in the traditional sense of a cavity or opening. The adjective "black" is used because, as light cannot escape from a black hole, it is completely dark, emitting nor reflecting light, making it "black" to any observer.

When certain stars, much more massive than the Sun, reach the end of their lives, they can undergo a process called gravitational collapse. After consuming all nuclear fuel, the pressure that supports the star against gravity disappears, and it collapses upon itself. Depending on the original mass of the star, this collapse can result in a supernova, and the remaining core can form a stellar black hole. This is an example of a black hole that originates from a "dead star." In this sense, we advance to the theory of a dead universe with dimensions larger than our observable universe.

If the Sun were to die, that is, if it suddenly stopped emitting light and heat, the consequences would be dramatic, but the orbits of the planets in the solar system, including Earth, would initially remain unchanged, at least for a certain time. This is because gravity, not light, is the force that keeps the planets in orbit around the Sun.

Gravity is a consequence of an object's mass, and light is a form of energy emitted by it. If the Sun suddenly stopped emitting light, it would mean that it is no longer undergoing nuclear reactions in its core, but its mass would still be present, and therefore, its gravity would continue to influence the planets. However, the absence of light and heat would have catastrophic effects on life on Earth and the planet's climatic conditions.

Over time, if the Sun were to turn into a white dwarf star or undergo some other process that significantly altered its mass, the orbits of the planets could be affected. Changes in the mass of the Sun would alter its gravitational force, which in turn would affect the trajectory of celestial bodies orbiting around it.

It is not strange to postulate the existence of a universe without the activity of light emission but still composed of galaxies, supermassive black holes, dark matter, dark energy, and where the laws of physics remain in force. I can affirm, based on the theoretical argument developed, that such a universe exists and will soon be revealed in the light of scientific knowledge.

From a scientific point of view, however, the assertion of the existence of a reality as fundamental as that of a 'dead universe' requires a substantial set of empirical and theoretical evidence, capable of being proved through independent observations and experiments. Until such evidence is provided and validated by the scrutiny of the scientific community, such a concept must be considered with caution, currently situated in the domain of theoretical speculation, akin to many hypotheses and theories that preceded it.

These black holes are found at the centers of almost all large galaxies, including our Milky Way. They have masses ranging from millions to billions of times the mass of the Sun. It is believed that they grow by accumulating matter and other black holes over time, but their exact origin is still a subject of research. They are not considered "dead galaxies," but they are a fundamental part of the dynamics and evolution of galaxies.

The term "dead galaxy" usually refers to a galaxy that has stopped forming stars. Galaxies can "die" in terms of star formation due to various processes, such as gas loss (the fuel for star formation) or interactions with other galaxies. These galaxies do not transform into black holes, although they may harbor supermassive black holes at their centers. The theory of the dead universe is legitimized and worthy of study by proposing that the deaths of celestial bodies converge in the formation of a singular predecessor universe, as opposed to the conception of multiverses suggested by various speculative theories. These theories often depart from rigorously tested and proven mathematical models and scientific evidence. In contrast, the theory of the dead universe harmonizes with established physical laws and offers alternative explanations that can be easily subjected to verification through observation and experimentation.

In the next article in partnership with astrophysicists, my purpose is to advance the presentation of a consistent model that, to achieve validation by the scientific community, must be able to make testable predictions and be robustly grounded in demonstrations and existing empirical data that favor the theory. Currently, the consensus around the big bang theory seems to be weakening, while, on the other hand, the theory of the "dead universe" not only conforms to the already established physical laws but also proposes alternative explanations that can be easily subjected to verification through observation and experimentation.

Thus, the "expansion of the universe" can be interpreted not as an indicative of dynamic growth, but rather as a gradual separation driven by the laws of gravity of a predecessor universe, a relic still influencing the current cosmos. This phenomenon could be considered as the final exhale of a universe that gradually yields its energies. We witness a process of cooling and tranquility, where matter and energy redistribute gently, and space-time stretches, aspiring to a state of lasting serenity. As this process continues, the formation of new galaxies will tend to decrease and eventually cease, resulting in a universe filled with contemplative silence—the true quietness that follows the luminous interlude of the stars. Just as its parent universe died, so its child, the observable universe, fades away . . .

Therefore, the legacy of the "dead universe" constitutes the key to understanding our cosmic destiny, focusing not on active galaxies but on the contemplation of the most ancient structures and the careful observation of celestial phenomena such as black holes. These investigations can unveil crucial clues about the primordial universe and provide a more comprehensive understanding of its beginning and conclusion, without resorting to repetitive cycles. The firmament that stretches beyond the known stars and galaxies is not a vacuum devoid of existence but rather a vastness filled with supermassive black holes in constant mergers, a universe where the most complete absence of light reigns, planets submerged in darkness, and where dark matter predominates with incomparable density, suggesting even the presence of particles unknown to our visible universe. There is undoubtedly a cosmos waiting to be deciphered, shrouded in mystery and which, in time, will be revealed before advanced instruments such as the James Webb Space Telescope and forthcoming technologies. The existence of supermassive bodies whose dimensions exceed by tens of billions those of the largest entities ever cataloged, and whose gravitation molds entirely inert galaxies hidden in the shadows, is in perfect accordance with the laws that dictate the mechanics of this universe even in the complete absence of light. According to the theory of the "dead universe," light or its absence does not constitute the determining criterion in characterizing a universe. The advent of such understanding, which challenges the notion that our universe has an age limited to 13.5 billion years, suggests that we must prepare for a paradigmatic adjustment, for undoubtedly, we have a scant interval to recognize that our previous conceptions may have been mistaken for a long period.

As we walk towards truth, we need to detach ourselves from less comprehensive theories like the big bang, which, although predictive, now gives way to a simpler and more enlightening model. The theory of the "dead universe" stands out for its clarity and the way it rationalizes observational data, offering a straightforward perspective on the empirical evidence pointing to a universe characterized by a single genesis followed by a definitive conclusion. We are on the right path but are embracing the wrong theories.

Many black holes are nothing more than tombs of new galaxies, just as our universe contracts within the vast abyss of an immense black hole from the dead universe. Therefore, this provides a response to the large amount of dark matter that surrounds our observable universe. We exist within a great cosmic tomb, and when we die as a universe, our funeral and burial have already been provided by the old, deceased universe.

As intrinsic inhabitants of this cosmic tomb, we are witnesses to our own final abode, already lodged within its confines. We are not heading towards this grim destination; we are already immersed in it. Thus, when our universe succumbs, we will indeed be within our own sepulcher, provided by the deceased universe that preceded us.

Black holes, often conceived as catacombs of nascent galaxies, illustrate the decline of our cosmos towards the vast vacuum of a colossal black hole, a vestige of the preceding universe. This perspective provides a revealing interpretation for the enigmatic proliferation of dark matter that permeates the visible universe. We reside, therefore, in the midst of a stellar cosmic sepulcher; and when the time comes for our universe to succumb, its epilogue will be preordained by the already consummated demise of the ancient dead universe.

Focusing on the study of the dead universe seems more sensible to me than seeking any evidence of extraterrestrial life, intelligence, or even distant galaxies. We should dedicate our efforts, resources, and energy to investigating the previous death, thus learning more about annihilation, the inevitable end of what once was our beginning and now heads towards the end of the end.

The newly discovered galaxies by the James Webb Space Telescope can be seen as the last echoes of an observable universe that is, in essence, the luminous vestige of an extinct cosmos. These galaxies are the heirs of a comprehensive stellar legacy, mere final whispers of a once expansive and energetic universe. We live, thus, under the shadow of an ancestral cosmic splendor, witnessing what may well be considered the twilight of the interaction between light and matter—the resplendent epilogue of a declining universe. Our current stellar reality is nothing but a distant echo—a delicate yet still resonant fragment of a reality much broader than the known boundaries of time and space. At the core of our existence, everything that exists, everything we observe, and everything within our reach of understanding are just the preserved remains of a larger universe that has vanished, the enduring signature of a dead universe.

Just as stars are born from clouds of dust and gas, our universe may have been formed in part from the "dust" left by its predecessor, but also the younger celestial bodies are objects that were born as the ancient universe was dying. That is, it created new galaxies with the signature of future death, even when it was dying. This process can be observed in our observable universe, following the laws of conservation and transformation that govern all maturity.

The universe is certainly slowing down, despite any theory suggesting its continuous expansion. Even the theory of the "dead universe" suggests that galaxies may drift apart. On the other hand, we will never see the emergence of a galaxy larger than those we know today in our own universe. This should be reason enough for the scientific community to take this theory more seriously, rather than just looking for flaws in the theory.

A universe dating model based on the big bang theory would be comparable to looking for a live dinosaur to explain its origins and lifespan. The truth is that we will only reach a consensus through the study of the dead universe. In this way, we will set precedents for understanding a universe that may have existed for trillions of years, where our 13.5 billion years represent no more than an insignificant fraction. We are just particles wandering in a space-time that was once almost infinite in its magnitude.

I am addressing observable phenomena, such as black holes, and I do not see my proposition about the dead universe as more speculative than the theories ventured by notable scientists, such as Lawrence Krauss regarding dark matter when it was still considered an unimaginable conjecture. Or even the various theories proposed by Albert Einstein and Stephen Hawking, which were proven many years after their initial formulation.

Therefore, considering the existence of the dead universe as lacking evidence seems more absurd than any criticism that could have been directed at these pioneering scientists. Furthermore, the theory of the dead universe already finds support in new data, including those provided by James Webb regarding supermassive black holes. Perhaps it is time to look through the rearview mirror of the big bang theory to discover what is emerging behind it, revealing the mysteries of the universe to the eyes of the scientific community.

Therefore, this theory could offer an alternative explanation for the origin of the universe and its subsequent developments. Asserting that the universe has always existed would be similar to saying that God has always existed; however, these primordial black holes, originating from the death of the predecessor universe, could vary greatly in size, from very small to extremely large. They are an active area of research, especially in their possible contribution to dark matter. As stellar or intermediate black holes interact with each other in binary systems or in dense regions of stars, such as galaxy centers or star clusters, they can collide and merge, forming a more massive black hole. These mergers are now regularly detected through gravitational waves, a form of radiation emitted by merging black holes. In fact, to some extent, the theory of the expanding universe finds better support in the theory of the "dead universe" than in the big bang theory, since young stars, in this process of the "dead universe," are heading towards death. So, we can think that the known universe is just 'living sparks' of the dead universe, which must also die in billions of years and are within the center of a black hole of the dead universe, like a womb with unknown dimensions of dark matter originated from the death and mergers of black holes of exorbitant dimensions.

The theory of the "dead universe" does not advocate for the eternity of the universe or argue that the universe is the result of an endless cosmic cycle. On the contrary, it proposes that our universe is the product of a previous universe, marking an endpoint in its existence, rather than a history of infinite creation and recreation as proposed in some theories of cyclical cosmology. This approach not only challenges the notion of uncritical acceptance but also presents itself as a straightforward and understandable explanation for the origin of the universe.

Although the big bang theory is widely accepted by the scientific community due to its precise predictions and agreement with observations, the theory of the "dead universe" aligns with and potentially surpasses the big bang in terms of explaining the origin of the universe with a simplicity comparable to creationist conceptions, but without relying on unscientific assumptions.

This alternative theory not only fits well with the evidence supporting the big bang but could offer even more precise and testable predictions. The theory of the "dead universe" could then become a crucial field of study in theoretical cosmology, challenging and possibly replacing the big bang paradigm as the main explanation for the origin and evolution of the cosmos. Such a proposal has the potential to revitalize scientific debate and offer new directions for the study of cosmology and astrophysics.

The theory of the "dead universe," which suggests a cosmos devoid of light activity but fully active in its origin, may have been born in a state of advanced maturity, unimaginably vast and filled with energies like dark energy itself that influences our universe. Over incalculable ages, however, this ancestral universe entered a phase of decline, where each subsequent cycle of stellar life appeared of lesser magnitude and duration than the previous one, always heading towards death.

In this context, the formation of new galaxies and stars, such as those observed by modern technology, does not signal the vibrant birth of an expanding universe, but rather the last gasps of vitality of an aging cosmic structure that yearns for life even as it approaches death. Our universe, at the moment of its birth, held an advanced state of development intertwined with signs of cosmic old age.

As it ages, instead of expanding and growing in complexity and diversity, the universe is paradoxically rejuvenating in terms of its galaxies and stars, which emerge increasingly young and smaller, denoting a progressive loss of energy and mass that will lead it to total death in two hundred billion years.

The "dead universe theory" is built on a solid foundation of observational data and established mathematical models. Although it does not make specific predictions now, it aligns with a series of scientific evidence that could be interpreted as congruent with its central postulates. The theory is proposed as a coherent and rational framework, offering a new perspective on the cosmos that is consistent with current physical and astronomical knowledge. The history of particle physics, particularly the prediction and subsequent discovery of the Higgs particle, highlights the patience required for scientific advancement. Important theorizations often precede the experimental verification capability by many years, if not decades. In this spirit, the "dead universe theory" stands as an invitation to continuous investigation, awaiting the development of technologies and methodologies that may, in the future, test its premises and enrich our understanding of the universe.

Under the theory of the "dead universe," the accelerated expansion of the universe is a natural phenomenon resulting from the immense energy released by the destruction of the old cosmos and the subsequent formation of the new. Dark energy, then, could be seen as an energetic vestige of this cosmic transition, a residual force driving the continuous expansion of the universe. The theory requires a reinterpretation of existing cosmological data, including cosmic microwave background radiation and galaxy distribution. For example, variations or irregularities in background radiation can be interpreted as evidence of a transition event between the dead universe and the new one.

To support the theory, advanced computational models can simulate scenarios of black hole collisions and the subsequent formation of a new universe. These models would help better understand how the described events could manifest in the observed structure of the current universe.

Black holes, long theoretically predicted, have only recently had their empirical existence consolidated. The theory of the "dead universe" can explain phenomena that the big bang theory may never be able to, such as certain characteristics of black holes and other cosmic phenomena. This theory has accepted, clear, and verifiable foundations that not only align with observations made from the perspective of the big bang but also predict phenomena that the big bang cannot efficiently explain. These collision events could be responsible for the anomalies observed in the cosmic microwave background radiation, which the big bang theory only partially explains. In the context of the "dead universe," these would be remnants of the last intense gravitational interactions of the previous cosmos.

Dark matter, an essential component of the cosmos that the big bang does not completely explain, finds a place in this theory as a direct remnant of the previous universe. The theory of the "dead universe" suggests that dark matter is composed of particles or compact objects that are remnants of the collapse of the ancient universe. Now the interpretation of dark matter provides a new angle to investigate its properties, as its distribution and behavior may reveal more about the conditions of the pre-existing cosmos than our observable universe.

The acceptance of a theory by the scientific community is not only a matter of accumulating evidence but also of paradigm shift. The history of science is filled with widely accepted theories that were eventually supplanted by new theories that provided more precise or comprehensive explanations. The theory of the 'dead universe' proposes a reinterpretation of already known phenomena and the possibility of more adequately explaining observations such as cosmic microwave background radiation, the abundance of light elements, and the accelerated expansion of the universe.

It is crucial that the theory of the "dead universe" be debated, tested, and potentially validated by the scientific community. This debate will not only contribute to the advancement of knowledge but also challenge the foundations of established theories, promoting a deeper and integrated understanding of the universe.

The substantial presence of dark matter in the universe suggests the validity of the theory of the "dead universe." Although the big bang is recognized for explaining the cosmic microwave background radiation, it fails to precisely determine the age of the universe. Astrophysics and cosmology, by basing the dating of the universe on still-active celestial bodies, propose that the universe is approximately 13.5 billion years old; however, this conception is destined for revisions in the light of the theory of the "dead universe." The new methodology suggested by advances such as the James Webb Space Telescope indicates that the observation of extinct stars may point to a much older age, possibly in the trillions of years.

This approach challenges the interpretation of gravitational waves within the big bang paradigm. In the theory of the "dead universe," the existence of an astronomical number of extinct stars in a chaotic and random universe, where collisions are frequent, offers a more plausible explanation for the gravitational waves detected near Earth. General relativity, therefore, strengthens the theory of the "dead universe," presenting a divergent perspective on the expansion of the universe.

In the view of the "dead universe," the visible universe consists of young galaxies emerging from the death of a precursor universe, driven by intense conflicts and collisions, phenomena hitherto unexplained by black holes, as observed.

The conception that our young universe represents the ultimate gasp of a past cosmos—a "dead universe" composed of trillions of galaxies and quantities of energy that challenge our quantification capacity—finds support in the latest observations from the James Webb Space Telescope. These observations point to the birth of still-young galaxies, emerging from the last energetic pulsations of a universe that, though dying, is still capable of generating new celestial structures.

The existence of these galaxies, propelled aggressively by a cataclysmic past, challenges timelines based on the big bang, which assumes a uniform expansion from a single point.

At the intersection of geology and astrophysics, we find divergent dating methods: while geology offers robust dating techniques through analyses of terrestrial residues, astrophysics and cosmology continue to explore the age of the universe through observations of active celestial bodies. This discrepancy underscores the importance of methodological revision. Recent observations made by the James Webb Space Telescope suggest that, in the vastness of the cosmos, the existence of our cosmic singularity aligns more with the theory of a "dead universe" than with the idea of a universal expansion originating from a singular point. Our universe, seemingly effervescent in the generation of new galaxies, may not be an independent and expansive entity, but rather a mere reflection, a diminutive and nostalgic simulacrum of the fullness of a pre-existing universe. Here, each new stellar and galactic formation is a replica, an echo of the memory and mechanics of a once full cosmos, now dissipated in its magnificence and dead.

The perspective that our universe may be the remnant of an ancient cosmos, extensive in time and space, calls for a revision of traditional narratives about cosmic origin and evolution. According to recent observations from the James Webb Space Telescope, we are presented with nascent galaxies forming from the remnants of a "dead universe," suggesting a direct inheritance from a stellar domain of trillions of years, brimming with energy and galaxies beyond what we can currently quantify.

This understanding suggests that the continuous formation of galaxies is not indicative of an infinite cyclical process of cosmic births and deaths, but rather a singular occurrence subsequent to the death of a primordial universe. The appearance of new stellar agglomerations is not simply an act of repetition, but the transformation of an ancient universe, marking a new phase in the cosmological continuum.

The indications of vibrant and young galaxies discovered by the James Webb telescope may be considered the most concrete vestiges of the ancestral universe, whose fundamental elements transfigure and give rise to new celestial bodies. These discoveries not only serve to confirm the diversification of the cosmos; they also represent a valuable document of stellar lineage, a narrative of the deeply rooted and intricate cosmological past.

In the face of these new understandings, it is urgent to revise and adapt our cosmological models to encompass and reflect upon the concept that the end of an ancient universe does not represent a conclusion, but rather the beginning of a new galactic generation. This renewed paradigm may pave the way for a new frontier of astrophysical discoveries, replacing the notion of finality with that of continuous transformation.

The integration of these concepts into the theoretical framework of the "dead universe" will not only enrich contemporary scientific debate but also provide a robust foundation for future investigations aspiring to elucidate the mysteries of the origins, evolution, and ultimate fate of the cosmos.

This theory presents an innovative perspective on the genesis and dynamics of the cosmos, in which the universe we perceive is shaped not only by its own substance and history but also by the reminiscences and gravity of a predecessor universe. Here, creation does not arise from a singular inflationary event like the big bang but emerges from the silence of an ancient and already disappeared cosmos. This preexisting cosmos, though absent in direct manifestation, inscribes its laws and structures into the fabric of our own universe, influencing both its expansion and its mass distribution. Thus, the universe we inhabit is not a creation from nothing as proposed by Lawrence Krauss but a continuation—a posthumous universe that carries the gravitational and structural legacy of a previous reality on a smaller scale, deeply intertwined in our cosmic existence and evolution.

The beauty of this theory lies in its ability to offer verifiable predictions. For example, it suggests that certain anomalies in universal expansion are normal and can be explained by the influence of the previous universe, offering a new field of study for astronomical observations. Detailed analyses of cosmic background radiation or dark matter distribution could reveal unexpected patterns, serving as empirical evidence for the dead universe theory.

It can be speculated that the dead universe with its constant mergers released a field of energy and dark matter that permeates the space of the current universe and affects and provokes the galaxies' distancing. If the previous universe underwent gravitational collapse or another extreme phenomenon, it caused residual quantum or relativistic effects that can influence the metric of spacetime of the remaining active universe, altering how galaxies drift apart from each other.

The foundation of this theory can be supported by the modification of the Hubble Law, traditionally expressed by the equation $v = H_o D$, where $v$ represents the velocity at which galaxies recede, $D$ is the distance to these galaxies, and $H_o$ is the Hubble constant. We propose an expansion of this law to incorporate the impact of the predecessor universe, introducing an influencing factor, $F(\text{dead universe})$, that adjusts the Hubble constant based on the properties of the "dead universe."

The influence of the dead universe is captured by the function F(dead universe) = $\alpha\rho$res + $\beta$C, where $\alpha$ and $\beta$ are constants that translate the relationship between the residual density ($\rho$res) of the ancient universe and other final conditions (C), such as gravitational or quantum residual effects. Thus, the velocity of galaxies' recession in our universe is redefined as v = ($\alpha\rho$res + $\beta$C) · Ho · D, an equation that reflects the interaction between the current cosmos and the fabric of spacetime of the dead universe.

Despite observations indicating galaxies drifting apart from each other, I do not conceive the traditional expansion model but rather apply the same laws that fit better into the dead universe theory, remaining skeptical of the expansive model suggested by the big bang theory. The dead universe theory emerges not only as an alternative cosmological model but aspires to the elegance of a "theory of everything," a unifying conception that promises to intertwine all empirical evidence, scientific data, and calculations into a single explanatory fabric. It is proposed that the current universe, with its shimmering stars and spiral galaxies, is not an isolated system but a fragment, the smallest remaining fraction of a much vaster antecedent cosmos. This preexisting universe, now in a state of cosmic twilight, has left us as heirs to its last vibrant portions of complexity and order.

Like the residual cells of a once vibrant organism, our universe contains within itself the fundamental information, the intrinsic memory of the larger body of which it once was a part. In the structure of each subatomic particle, in the curvature of spacetime, and in the orchestrated movements of constellations, we find the echo of this majestic origin. Everything we consider to be natural laws, constants, and fundamental variables may be the reflection of the eternal dynamics of this "dead universe," with its laws still whispering through the expansion of space, guiding our expansion and our gravitation, our light, and our darkness.

This perspective suggests that what we seek to understand about our universe—dark matter and energy, the quantum nature of reality, the very fabric of the cosmos—are residual characteristics, preserved aspects of a larger and more comprehensive reality. Thus, in the quest to comprehend the origin and destiny of our universe, we might actually be deciphering the legacy of what we once were: a complete cosmos, now merely whispering the secrets of its past existence in the shadow of its own stellar demise.

Grounded in the theory of the dead universe, I venture, specula-tively, the possibility that the universe is, in fact, condensing toward a singular point of death with a new model of galactic separation as it hap-pened before, expelling us from the womb of the dead universe. The dead universe theory could shed light on the existence of supermassive objects far removed from any previous reality, as well as radio waves, echoes, dark matter, dark energy, and even reveal that UNO and still 'invisible' matter permeates the known cosmos.

My idea of addressing the initial subject of this work on the origins of the universe, based on a careful analysis of Gen 1:1–2, was precisely to set precedents for the consistent approach to follow. I will raise the ques-tion of "materia UNO," from the Greek (ἕνα), which represents undetect-able properties. We cannot conclude that it would be a total absence of matter, but rather that it cannot be detected due to possessing properties strange to our technologies. This «materia Uno" would be the connecting element of all particles, perhaps the vacuum. I intend to use the strength of this term for the absence of matter, but emphasizing that it exists as an essential element, emerging from the vacuum as a motor of antiparticles.

In the scientific context, the idea of Uno matter can be associated with various theories and speculations about particles or substances that may exist beyond the reach of current observation tools. For example, in particle physics, there are hypotheses about the existence of exotic sub-atomic particles, such as sterile neutrinos, which have extremely weak interactions with ordinary matter and are therefore difficult to detect.

Another possibility is that Uno matter is related to concepts in theo-retical physics, such as dark matter or dark energy, which make up the majority of the observable universe but whose exact nature is still un-known. These forms of matter may be present in significant quantities in the cosmos; however, their lack of interaction with light and other forms of radiation makes direct detection challenging.

At a more speculative level, Uno matter can be conceived as a form of matter that exists in additional dimensions beyond the three spatial di-mensions and one temporal dimension that we perceive in our everyday universe. Theories such as string theory and loop quantum gravity posit the existence of extra dimensions, where new forms of matter and energy may reside, thus escaping direct detection.

Both Lawrence Krauss and Stephen Hawking are known for explor-ing the idea that the universe may have originated "from nothing" or from a state of quantum vacuum. Their perspectives on this subject are quite similar in some respects, but they also have subtle differences.

In his exploration of modern physics, Lawrence Krauss discusses how quantum mechanics reveals that spaces considered "empty" are not truly void but teem with activity that can lead to the creation of particles and antiparticles. His analysis suggests that the universe could potentially have arisen from these quantum fluctuations. This understanding comes from the broader discussions in his book about the dynamic nature of the vacuum in quantum field theory and its implications for the origins of the universe.

Stephen Hawking, on the other hand, also discussed the possibility of the universe coming from nothing in his book *The Grand Design*, co-written with Leonard Mlodinow. He suggests that, due to the laws of gravity and quantum mechanics, the universe may have spontaneously arisen without the need for an external cause. Hawking argues that, given the highly compressed and hot state at the beginning of the universe, the laws of physics would allow it to emerge "from nothing" as a singularity.

Both scientists are dealing with the complex concept of "nothing" in a somewhat different way than is commonly understood. Instead of "nothing" meaning a true absence of anything, they are referring to a state of quantum vacuum that, although empty of matter and energy as we know it, is rich in quantum fluctuations and potentially capable of generating entire universes. Their ideas challenge more traditional conceptions of the creation and origin of the universe and continue to be subjects of debate and investigation within the scientific community.

As we contemplate the cosmic microwave background, we observe the primordial remnant of the inaugural luminosity of the cosmos, which permeates the universe uniformly—a true cosmic fossil. This evidence harmoniously aligns with the dead universe theory I am proposing. It postulates that the CMB is the indelible impression of a previous stage of the universe, an era of quiescence or equilibrium before a significant cosmic transition. The neutrality of the first atoms metaphorically reflects the universe in a state of latency or "death," a phase preceding the complexity and structuring observed in the present. This speculative theory suggests that we live only a phase in a much broader and perhaps eternal cosmic cycle, where the universe oscillates between periods of activity and inactivity.

At the heart of the unfathomable cosmos, we dare to hypothesize beyond the scope of current scientific consensus: it is possible that our universe, with its majestic expansion and intricate complexity, far exceeds the age estimated by our contemporary methods, reaching or even surpassing the one hundred billion-year mark. This bold conjecture is based on the previous existence of a "dead universe," a cosmic structure whose longevity would have extended for a period exceeding nine hundred billion years. It is suggested that such an ancient domain, ending its extensive cycle of existence, would have been the precursor to our current universe, originating through a phenomenon of rebirth or cosmic metamorphosis. The conception that we inhabit a universe succeeding an even older and more expansive reality not only broadens the horizons of our understanding of the temporal dimension of space but also inaugurates new paths of inquiry into the cosmic lifecycle and the primordial laws governing its trajectory. Although this theory resides on the fringes of scientific speculation, it invites thinkers to ponder the truth of a space-time whose fabric and matter may represent not the beginning, but a more recent phase of a universal cycle of immeasurable scope and ancestral reverberations.

Despite the assertion of astronomers and cosmologists, I rely on an elementary question: what is the true age of supermassive bodies? What changes with a dating method that takes this premise into account? Should the chronology of the universe be measured exclusively by the lifespan of stars? If we are mistaken in this methodological model, how then should we interpret the recent findings of the James Webb Space Telescope? The newly observed galaxies are notably vast and contain stars whose redshift signals an ancient provenance, dating from a mere five hundred to seven hundred million years after the event known as the big bang. At the dawn of the cosmos, we observe that black holes already had immense masses, much larger than that of our Sun, a finding that challenges current explanations, suggesting that they may have formed before the initial event proposed by conventional cosmology. It's not that celestial phenomena are inherently inexplicable by the laws of traditional physics, which cannot be used in the future for universe dating questions.

Therefore, chronometry based on galaxies and stars, whose existence is finite, may not be the most accurate method for dating the extent of cosmic history. By analogy, the 'age' of the universe could be inferred by studying the "fossils" of a previous cosmos, much like how the age of a living descendant can be contextualized through the study of the remains of the progenitor. Investigation of these cosmic remnants may reveal not only the duration of the past existence of the deceased "parent" but also provide insights into the potential longevity of the "child"—the observable universe.

This realization coherently aligns with the theory of the dead universe, from which these colossal structures would have emanated. Faced with the bewildering massiveness of black holes, their reconciliation with the theory of the primordial explosion and the subsequent expansion of the universe is questioned, often illustrated as an "inflating balloon." How can we incorporate such colossal entities into current cosmological equations? It seems plausible that such black holes are not mere by-products of our emerging universe, but rather remnants of a previous cosmos, existing long before the coalescence of galaxies that outlined the cradle of our universe.

Based on emerging astronomical observations, my assertion maintains that the age of our universe may well exceed one hundred billion years, emerging not from the dawn of a new reality, but from the decline of a past universe. This premise is not unfounded but anchored in data pointing to an older and more complex origin. If we accept that less massive galaxies have the potential to exist for approximately one trillion years, then why should we discard the hypothesis of a universe with more than one hundred billion years as mere fiction?

In an immeasurable universe, I contemplate the possibility that the cosmos in which we reside is the descendant of a previous universe. A dead universe. The colossal stars of this ancient domain, billions of times larger than the largest we know, would have lived and died, with their deaths sowing the seeds of a new beginning. As these giants succumbed, the celestial bodies orbiting around them would have been drawn to a central birth point—the womb of our own universe. The residual radiation, like cosmic radio waves, would be the signature of this colossal process, concentrating around our nascent universe, which is filled with noticeably smaller stars. This narrative is reinforced by observations from the James Webb Space Telescope, which reveal immense and mature galaxies prematurely in the young universe, suggesting a cosmic cycle of death and rebirth. I am skeptical of a cosmos limited to a mere 13.8 billion years, and "nurture" the conviction that the divine inaugurated his creation by the end, orchestrating all existence with a transcendental design from its inception.

As evidence emerges of the existence of supermassive bodies and various types of black holes, demonstrating a universe that exists in a state different from the observable universe, this theory is consolidated.

While the big bang fails to explain these phenomena, it gives way to a new theory. If correct, the presence of dark matter in the observable universe would be a strange element, not belonging to this universe but present as a residue of the previous universe, as well as dark energy itself. The constant mergers and residual phenomena of this dead universe would directly influence the observable universe, providing more objective explanations for the equations of general relativity theory and inexplicable phenomena of quantum mechanics, considering the existence of a neutrality, which would be the matter supporting the observable universe in conflict with elements that should not be present in our universe.

Dark energy emerges as one of the most fascinating enigmas of contemporary cosmology, originally introduced to elucidate the remarkable observation that the expansion of the universe is accelerating—a discovery made in 1998 through the study of the brightness of distant supernovae. The pioneering research of Saul Perlmutter, Brian P. Schmidt, and Adam G. Riess, deserving of the Nobel Prize in Physics in 2011, consolidated the acceptance of dark energy as a vital component in the current cosmological description. However, contrasting with this paradigm, the theory of the "dead universe" offers an alternative explanation for the galaxies' recession that does not require continuous expansion, attributed by conventional theory to dark energy. According to the theory of the "dead universe," what is perceived as dark energy could be interpreted as remnants of a previous cosmos, and dark matter, a relic of that antecedent cosmic death. In this perspective, the presumed dark energy does not play a role in the accelerated expansion of the universe; galaxies recede under the residual influence of physical laws of a "dead universe." Thus, the presence of dark energy, instead of indicating accelerated expansion, aligns with possible evidence corroborating the existence of the "dead universe."

Recent theoretical propositions have challenged our understanding of the cosmos and its genesis, among which stands out the hypothesis of the universe as an information processor, conceived by scholars at the prestigious University of Oxford. This approach suggests that the universe operates as a complex system of informational exchange, where each element, from subatomic particles to celestial bodies, actively participates in a dynamic network of data exchange.

Parallelly, the theory of the "dead universe" resonates harmoniously with this view, postulating that there are informational interactions between the deceased cosmos and the universe in which we are situated. This ongoing dialogue between what was and what is presents an alternative paradigm to the traditional big bang model, contemplating the possibility that the fabric of space-time is permeated by a constant flow of information from a preceding cosmological reality.

Moreover, everyday objects, such as chairs and computers, are considered participants in this gigantic cosmic processor, integrated into a universal informational matrix. In contrast to more complex models, the theory of the "dead universe" offers an elegant and explanatory simplification of the cosmos, a system that may be, in essence, a vast archive of perpetually interconnected and timeless information.

A nuclear physicist poses an intriguing question: how is it possible for an empty atom to form the ground that surrounds us? For a long time, it was not known that the atom was largely empty. It was only with the advancement of science, especially in the study of physics and quantum mechanics, that this discovery was made. Projects such as the LHC (Large Hadron Collider) were developed with the aim of exploring these questions and seeking answers. One of these questions led to the discovery of the Higgs particle, fundamental to our understanding of the origin of all things.

The James Webb Space Telescope, now in orbit, is truly a remarkable feat in the quest for knowledge of the cosmos. This epic endeavor not only recalls the adventures seen in space films but also reflects the hard work and collaboration of over ten thousand people who came together to launch it into space. This mission represents an extraordinary challenge, as the equipment is incredibly sensitive, presenting three hundred and forty-four potential points of failure.

Exploring the "observable universe" is essential for deepening our understanding of the mysteries of the dead universe, as this truth will be discovered unprecedentedly. The term "observable universe" refers to the part of the cosmos that we can directly observe, whether through terrestrial or space telescopes. This region encompasses everything we can detect through light and other forms of electromagnetic radiation. The observable universe includes stars, galaxies, nebulae, and other celestial bodies that emit light or are illuminated by light sources. However, this observable portion of the cosmos represents only a small fraction of the total universe that is actually dead, as there are vast regions inaccessible to direct observation due to distance, darkness, or other factors.

Several scientists have discussed the idea that the universe arose from nothing, using different definitions of what that means in a physical and philosophical context. Lawrence Krauss is one of the most well-known proponents of this idea, which he details in his book *A Universe from Nothing*. He explores the notion that the universe could have arisen from a state of potentiality, where "nothing" is not an absolute absence of everything, but a state where the sums of all energies of the universe could result in zero, making the emergence of the universe a physical possibility.

Stephen Hawking also ventured into discussions about the origin of the universe. In his explorations of the complete theory of the universe, he suggested that if we could find a complete theory, it would allow us to participate in the discussion of why the universe exists and potentially know the "mind of God." He previously stated that the existence of a creator was not incompatible with science, although his later positions seemed to contradict that.

In attempting to understand the origin of the universe, scientists like Stephen Hawking and Lawrence Krauss explore concepts that echo ancient ideas, some of which find parallels in religious texts, such as the notion of creation *ex nihilo* ("creation out of nothing"), a concept present in various religious traditions, including the Bible. While these scientific perspectives are generally well-received in academic circles due to their empirical and theoretical basis, similar interpretations from theological sources are often viewed with skepticism or disregarded for not following traditional scientific methodology. This contrast in the reception of ideas highlights the complex interaction between faith and science and the importance of methodology in validating theories within the scientific community. However, the historical recognition that concepts of creation out of nothing exist in ancient texts opens an interesting dialogue about the evolution of human thought regarding the origins of the cosmos.

Black holes, far from being mere relics of collapsing stars, may represent residual phenomena from a past cosmic era, possibly acting as sentinels of distant cosmic events not yet fully understood. These enigmatic entities may hold clues to physical processes from a predecessor universe, challenging astronomers to decipher their history and contribution to the framework of the current cosmos. Contemplating the hypothetical origin of a dying universe, stretching over trillions of years and whose essence seems to have been transplanted into the present configuration of spacetime, leads us to reconsider the traditional narrative of the big bang. The scenario that emerges suggests that if a major explosion event occurred, its advent may have been much earlier than the chronology proposed by George Lemaître, prompting reflection on the temporal and structural complexity of the universe in which we reside.

As we delve deeper into the understanding of these celestial mysteries, we are led to speculate on what else may exist beyond the reach of our telescopes and measuring instruments. Black holes emerge as silent guardians of long-buried cosmic secrets, patiently awaiting the moment when science will fully reveal them, unraveling the mysteries of cosmic existence and our own origin in our small universe.

From this understanding, we can begin to explore the properties of black holes within the perspective of this theory, as there will be a dating for them as proposed. One of the most intriguing features is the event horizon, which is the boundary beyond which nothing can escape the gravitational attraction of the black hole, becoming understandable in the light of this theory. This horizon is a well-defined boundary in spacetime, and any object that crosses this limit is destined to fall into the black hole.

Another interesting aspect is the singularity at the center of the black hole, where the density and curvature of spacetime become infinite. This singularity is a point where the laws of physics as we currently know them cease to apply and is one of the great mysteries of theoretical physics.

The theory of the dead universe about black holes as gateways to an alternate reality of a dead universe shares parallels with the theoretical framework proposed by physicist Michio Kaku. Kaku's exploration, described in *Parallel Worlds: A Journey Through Creation, Higher Dimensions, and the Future of the Cosmos*, suggests the possibility of black holes serving as channels to other dimensions. In this scientific context, the connection between the theory of the dead universe and Kaku's ideas lies in the speculation that black holes could facilitate travel between different regions of spacetime, potentially providing access to alternative dimensions within a broader cosmological framework. This conceptual alignment underscores the importance of investigating black holes as potential gateways to unravel the mysteries of the universe's structure and composition.

I conjecture that the universe from which the visible universe we observe emerged has nearly infinite dimensions and an almost incalculable gravity, which would explain its ability to influence and curve spacetime along some known axis in our visible universe.

At the forefront of cosmological research, the study of black holes reveals that we reside in a fraction of a universe of nearly limitless mass, filled with known matter, encompassed by a space of astronomically expanded pre-existing dimensions. The overwhelming gravitational influence, along with cataclysmic mergers of black holes and events yet to be elucidated, may have propelled active galaxies beyond the confines of a previous cosmos—a dead universe. This process, for reasons still uncertain, seems to have triggered a singularity, a new distinct entity, in which a smaller universe, albeit rich in observable phenomena such as life and light, emerged with dark energy and matter and physical laws remaining in force.

From this new perspective, we can deduce that the legacy of an extinct and obscure universe maintains its influence on the architecture of the cosmos we inhabit. The existence and complexity of black holes, as well as other yet unexplained astrophysical phenomena, may be indicative of this influential continuity, suggesting that the cosmic past persists in shaping the present reality.

This hypothesis offers a potential interpretation for the presence of dark matter and other entities not fully elucidated by traditional physics. The proposal suggests the existence of an underlying or superimposed structure to the known fabric of our universe, a dimension where light, as we know it, is not present. In the catastrophic scenario of a dying star, whether by explosion or by the cessation of its nuclear activity, it is proposed that the release of energy is of such magnitude that it could disturb the spatial structure of the known universe. This would result in the manifestation of a void, a gap that would provide a window into an unknown domain, possibly a remnant of a pre-existing universe devoid of luminosity.

My analysis suggests that the force to create a fold in spacetime, as proposed in this theory, may derive from the essence of the ancestral primordial universe, which, due to its exceptional density and gravity, distorts spacetime.

As we contemplate the ultramassive bodies and the almost inconceivable density of the ancestral universe, it becomes evident that the laws of gravity operating at these scales are not only intense but extraordinarily powerful. These forces not only curve spacetime but are capable of bending it to the point of radically transforming the structure and evolution of the cosmos as we know it. Such massive gravitational distortions could theoretically alter the rate of temporal flow, challenging our conventional understanding of causality and continuity in the universe.

The remnants of the dead universe in activity, where we live, communicate with the ancestral dead universe, where there must exist, in addition to the incomprehensible gravity by physics, also an incalculable concentration of dark matter, which we also attribute to the equation of the fold of spacetime in the observable universe, with the existence of light. A universe that coats another universe, but with powerful gravitational density, and an exorbitant layer of dark matter interacting with a universe where light exists and also with a cosmic fabric less dense than the ancestral universe, causing distortions in spacetime in our young universe of about 13.8 billion years.

If we consider the hypothesis of cellular memory, where cells beyond neurons are capable of retaining and transmitting memories and behaviors, we could establish an intriguing parallel with the cosmos. Just as transplanted organs carry echoes of past experiences to new bodies, perhaps our observable universe, in its genesis and evolution, is the manifestation of a "cosmic DNA" inherited from a dead universe. This current universe, filled with stars being born and galaxies in rotation, can be seen as a celestial body also in decline towards the end that, although distinct in form, continues to echo the "habits and tastes"—or the laws and mechanics—of its past existence. The new galaxies would be like acquired behaviors, remnants of a deep universal memory of the dead universe, indelible and perpetuated beyond the death of the ancient universe, supporting the notion that even in the depths of forgetfulness, the essence remains, guiding the rhythm of cosmic creation new galaxies. The theory of the dead universe postulates that the fabric of the cosmos that surrounds us is intrinsically marked by cosmic memories, recorded in the essence of every existing particle. This ancestral legacy could explain the seemingly bizarre quantum behavior of subatomic particles, which, in the depths of quantum mechanics, reveal interaction patterns that defy our conventional understanding of space and time. It could be conjectured that such particles, now distant from the harmony of a full universe, behave as if displaced, yearning for the intrinsic order of a larger and more complete cosmos, from which our observable universe is just a shadow or fragment. Thus, phenomena such as quantum entanglement and superposed probabilities may not only be fundamental characteristics of our universe but also echoes of a previous cosmic symphony, where the laws of physics operated on a scale of complexity and unity that now seem strange and unattainable.

This conjecture proposes that dark matter, along with other cosmic singularities, may actually be remnants of a primordial cosmos not yet mapped out, each one a latent evidence of the persistent influence of that original universe on the conditions of our current cosmos. Under the scrutiny of contemporary physics, the cataclysm of a star—whether by its thermal death or by supernova—releases colossal amounts of energy that, hypothetically, have the potential to break the boundaries of observable spacetime. This could create conduits to the dimensions of a "dead universe," providing a glimpse of the fundamental structures of the vast cosmos from which our reality emerged.

The James Webb Space Telescope, which has unveiled galaxies with unexpected attributes, the theory of the "dead universe" offers a new interpretation. Instead of a young and incipient universe, the observations can be seen as evidence of an older inheritance, a continuity of a previous cosmos. This alternative paradigm suggests that galaxies are not newly formed but may have evolved from an already established cosmic infrastructure, an inheritance from the universe that came before.

The dead universe theory, and the observations of James Webb, which revealed ancient galaxies with unexpected characteristics for the big bang model, make it clear the existence of the DEAD Universe FACTOR. According to this theory, the concept of a "dead universe" may offer an alternative explanation for the observation of galaxies and cosmic structures seemingly mature that exist in advanced states of development. Such formations, which exhibit unexpected complexities for their presumed age according to the standard big bang chronology, could be interpreted as remnants of a pre-existing cosmos. The assumption is that the conditions of an ancestral universe influenced the accelerated maturation of such systems, suggesting that the distribution and evolution of these galaxies may not be restricted to the temporal framework imposed by the big bang model, but possibly extending over a more extensive and intricate period, inherent to the deep past of the universe.

The theory of the "dead universe" seeks to coexist with the fundamental principles of physics. My theory of the dead universe proposes that the gravitational attraction of this previous universe, although not directly observable, shapes the fabric of spacetime in a manner compatible with the theory of general relativity. The influence of this primordial universe could be investigated as an underlying force that transcends the current understanding of quantum mechanics and gravity, challenging scientists to rethink the interaction between the large structures of the cosmos and the behavior of subatomic particles.

Based on the theory of the dead universe, I anticipate that future observations may unveil unexpected patterns in the distribution of dark matter and dark energy. These patterns could challenge the explanations offered by the big bang model, as my theory suggests an ancestral gravitational influence that still permeates the cosmos. I predict that black holes may be more than just the end of stars; they may act as channels to this primordial universe, revealing properties of spacetime that are distinct from what we know.

There is anticipation of the opportunity to observe galactic motion patterns that defy the expectations established by big bang projections. Such discrepancies may be attributed, according to my theory, to the residual gravitational influence of an ancestral universe, characterized by an almost unlimited density that surprisingly may still be impacting the dynamics of our observable universe. Furthermore, interpretations of redshift in distant galaxies may require revision in light of this concept, suggesting that universal expansion may be a more intricate and heterogeneous phenomenon than the mere isotropic expansion proposed in conventional models. Thus, the mathematical models that underpin the big bang theory could be questioned as new evidence and analyses corroborate the nuances presented by this alternative theory.

It is postulated that black hole mergers, along with meticulous analysis of spacetime curvature, could offer fundamental clues. Through the lens of my theory, black holes would not be mere gravitational anomalies, but rather luminous indicators that point us towards a deeper understanding of the cosmos. They could represent points of connection with the legacy of a universe that preceded us, an extinct entity whose darkness still permeates and shapes the foundations of our current cosmos.

The theory of the "dead universe" presents itself as an alternative and potentially more congruent conception to the phenomena described by general relativity and quantum mechanics, compared to the model established by the big bang. This theory advances the hypothesis that the extraordinary gravitational forces of a preceding cosmos may have been the shaping agents of spacetime in the universe we observe today. This would imply that the formulations of general relativity are not restricted to our visible cosmos but extend to encompass interaction with a prior and broader domain.

Moreover, the enigma of dark matter and energy could be elucidated within this theoretical framework as remnants of this "dead universe." Such remnants would not merely be floating in the vacuum of space but actively defining the structure and evolution of the visible cosmos. This could provide a new perspective for observing the accelerated expansion of the universe, or even offer clues to a possible future contraction, as well as explain the gravitational anomalies we have recorded. In summary, the theory of the "dead universe" has the potential to redefine our understanding of cosmic fabric and the very essence of gravity and universal dynamics.

The theory of the dead universe also connects with observations from the James Webb Space Telescope, which revealed ancient galaxies with unexpected characteristics for the big bang model, suggesting that this model is nearing its end as a reliable cosmological paradigm. In 2022, the telescope enabled the detection of an ultramassive black hole, thirty billion times the mass of the Sun, being the first to be measured using gravitational lenses. This method observes the attraction of a celestial object by the passage of light, providing strong evidence for the theory that there existed a previous, supermassive universe, whose amount of mass is incomprehensible, perhaps hundreds of billions of times larger than our known universe. The amount of energy, certainly, could be hundreds of billions of times greater than our universe, which lies within a small black hole in the womb of this immense dead universe.

The formation and nature of black holes remain one of the deepest mysteries within the context of modern cosmology, challenging the explanations provided by the big bang paradigm and the theory of relativity. Furthermore, the ubiquitous presence of dark matter may be more coherently addressed when considering the theory of the "dead universe," which postulates a preceding cosmos as the source of such phenomena.

Intriguingly, this theory finds a surprising resonance in ancient narratives, such as the account of Genesis, which, interpreted metaphorically, describes a process of formation and transformation of the cosmos. The passage alludes to a "recreation" of the universe, an idea that has intrigued both theological thinkers and scientists over the centuries (Gen 1:1–2).

The image of a universe emerging from disorder and obscurity, as described in sacred texts, can be seen as an allegory for a cosmic event of great magnitude—possibly a singularity or a primordial state preceding our current understanding of physics. While the big bang does not offer an explanation for a preceding existence or for the transition from "nothing" to "something," the idea of a cosmological recreation or rebirth echoes the notion of a universe that is more of a continuum than a singular and absolute origin.

Based on the theory of the "dead universe," it is proposed that the death of an ancestral cosmos, over trillions of years, was marked by the progressive production of dark matter, culminating in a force that directed newly formed galaxies towards a central nexus, known today as the observable universe. It is postulated that from this epicenter, containing approximately two hundred billion galaxies, emerged our current universe.

The recent observations of the James Webb Space Telescope corroborate with this notion, evidencing structures that can be interpreted as the "three pillars of creation" within this context. This theory provides an explanatory framework for the abundance of dark matter, suggesting that the phenomenon of universal expansion may, in fact, be a manifestation of a preceding matter concentration.

Additionally, it is conceivable that, in the decline of this primordial universe, cataclysmic explosions and hitherto unknown laws may have acted to coalesce galaxies, stars, and planets towards a singularity, possibly a supermassive black hole. The laws of gravity, within this new context, could be adjusting to the clustering of these massive celestial bodies.

Therefore, our universe may face a fate similar to that of its predecessor, either through continuous expansion or eventual contraction culminating in a new singularity. This process could occur on a temporal scale of less than one hundred trillion years.

Furthermore, the interaction between general relativity and subatomic particles of quantum mechanics may be influenced by remaining laws of the "dead universe." The existence of a form of matter thus far undetected, perhaps "UNO matter," could be responsible for suturing the fabric of our universe in order to maintain the integrity of the laws of physics currently observed.

Recognition must be given to the need to construct a substantial theoretical framework to lend scientific credibility to this hypothesis. Such a structure must rigorously describe the properties and dynamics of interaction of the postulated "UNO matter," elucidating its role in shaping spacetime and the origin of the universe. It is imperative to draw inspiration from advances in particle physics, notably the Standard Model and quantum field theory, which provide a deep understanding of elementary particles and the fundamental forces that orchestrate cosmic interactions. Deepening and expanding these paradigms may shed light on the underlying mechanisms that possibly govern the manifestations of the "dead universe," encouraging the scientific community to refine and test this theory with the necessary rigor for possible integration into the canon of contemporary cosmology.

Furthermore, it is clear to me that overcoming the challenges of detecting UNO matter will require significant technological advances. Just as the Large Hadron Collider was crucial in identifying the Higgs boson, we will need new instruments and experimental methods to explore UNO matter. This could mean the construction of even more advanced observatories, the conduct of experiments in high-energy physics not yet conceived, or even the development of revolutionary computational techniques.

The amount of dark matter, which I will call Mb, and the amount of UNO matter, which I will call Md, are crucial to understanding the nature of the universe. Furthermore, the intensity of interaction between dark matter and UNO matter, represented by I, plays a vital role in our calculations.

These equations indicate how the quantities of dark matter and UNO matter change over time due to their mutual interaction. The constant $I$ represents the intensity of this interaction, which may vary depending on the conditions of the system.

Furthermore, it is clear to me that overcoming the challenges of detecting UNO matter will require significant technological advances. Just as the Large Hadron Collider was crucial in identifying the Higgs boson, we will need new instruments and experimental methods to explore UNO matter. This could mean the construction of even more advanced observatories, the conduct of experiments in high-energy physics not yet conceived, or even the development of revolutionary computational techniques. The amount of dark matter, which I will call Md, and the amount of UNO matter, which I will call Mn, are crucial to understanding the nature of the universe. Furthermore, the intensity of interaction between dark matter and UNO matter, represented by $I$, plays a vital role in our calculations. These equations indicate how the quantities of dark matter and UNO matter change over time due to their mutual interaction. The constant $I$ represents the intensity of this interaction, which can vary depending on the conditions of the system.

Let:

- Mn be the quantity of UNO matter,
- Md be the quantity of dark matter, and

- *I* be the intensity of the interaction between UNO matter and dark matter.

The equations describing the dynamics of this interaction are:

$$\frac{dMn}{dt} = -I \times Md$$

$$\frac{dMd}{dt} = -I \times Mn$$

This simplified formulation captures the essence of the interaction between UNO matter and dark matter. However, a more in-depth analysis requires additional considerations about other forms of matter and energy, as well as equations describing the dynamics of the universe on a large scale.

Let:

- *Mn(t)* be the function describing the spatial and temporal distribution of UNO matter in the universe,
- *Md(t)* be the function describing the spatial and temporal distribution of dark matter in the universe, and
- *I(Mn,Md)* be the function representing the intensity of the interaction between UNO matter and dark matter, dependent on the local densities of both.

Assuming that the interaction between UNOl matter and dark matter is governed by a nonlinear differential equation, we can express the dynamics of the system as follows:

$$\frac{dMn}{dt} = -I(Mn, Md) \times Md$$

$$\frac{dMd}{dt} = -I(Mn, Md) \times Mn$$

These differential equations describe the temporal evolution of the distributions of UNO matter and dark matter in space. The function *I(Mn,Md)* encapsulates the complex physical interactions between these forms of matter.

Therefore, I see the search for UNO matter not as an escape into the realm of metaphysics, which I am absolutely against, but as an exciting and legitimate challenge for modern physics. This endeavor not only obliges us to deepen our understanding of the cosmos but also to innovate in the tools and theories that, I hope, will one day unravel its mysterious nature. I am convinced that we are on the eve of a new era of discoveries, where the shadows of the unknown will finally dissipate under the light of knowledge and technological innovation.

The existence of this UNO matter causes particles to behave differently in the face of unknown gravitational fields. The existence of two distinct entities of dark matter and undetectable UNO matter, originating from the demise of the previous universe, signals a transcendental influence on the dynamics of elementary particles and, by extension, on the phenomena that permeate the very essence of life. The detection and understanding of these elusive forms of matter constitute one of the most pressing enigmas of contemporary physics, evoking a cosmic dance that shapes not only the structure of the observable universe but also the intricate patterns that govern the very fabric of existence.

The "UNO invisible" matter as an antithesis to dark matter suggests a fundamental duality echoing the wave-particle duality principle in quantum mechanics. In this conception, it's possible to interpret the observed peculiarities in the behavior of subatomic particles, such as quantum leaps and the seemingly divisible nature of matter, as reflections of a mirrored reality between UNO matter and dark matter. From this perspective, the continuous interaction between these forms of matter offers an explanation for seemingly paradoxical phenomena, such as the dual behavior of light and the observation of interference in the double-slit experiment.

This approach suggests that the constant exchange of information between UNO matter and dark matter plays a fundamental role in structuring the fabric of the universe and in the manifestation of observed quantum phenomena. Furthermore, it's proposed that the interconnection between these entities transcends the traditional boundaries of classical physics, paving the way for a deeper understanding of the nature of reality and the fundamental laws that govern it.

Considering this perspective, we can envision a new approach to solving persistent puzzles in quantum mechanics and advancing toward a more complete understanding of the nature of the universe and our own existence within it.

These mysterious entities, by their very evasive nature, challenge the boundaries of human knowledge, suggesting the presence of hidden dimensions and fundamental laws that transcend our conventional conceptions. Their intrinsic role in shaping the cosmos and sustaining cosmic order sheds light on an intricate web of interconnections, where each particle, each galaxy, each manifestation of life is woven into a cosmic pattern of complexity and harmony.

As we venture into the abysses of space and time, we contemplate not only the distant past of the universe but also its uncertain future. The duality between expansion and concentration, between stellar birth and death, between light and darkness, confronts us with the cosmic imperative of incessant change and transformation. In this constantly flowing cosmic panorama, we are compelled to question not only our understanding of the universe but also our own existence and place within it.

Therefore, the investigation of dark matter and UNOl matter transcends the boundaries of conventional science, inviting us to explore the deepest mysteries of the cosmos and contemplate the most intimate mysteries of our own human condition. In our quest for understanding and meaning, we are guided by the promise of unraveling the secrets of the universe and, perhaps ultimately, the secrets of our own soul.

The theory of the dead universe better explains the theory of the "expansion of the universe" in light of Hubble's laws. We cannot in any way believe in an expansion from the explosion of the big bang that at some point is not explained; it makes no sense because indeed the universe would decelerate at some point in the billions of years of its existence. In fact, galaxies are moving away from each other in both theories, but not due to the expansion of a previous explosion, since there is not enough energy in the cosmos to cause continuous expansion. The gravity of the previous universe and facts of unknown laws, as explained in the gemstone question, are attracting the observable universe back into itself; this strong attraction explains many phenomena that were previously complex for quantum physics to explain. This theory will surely be elucidated in light of scientific evidence soon, as research focuses primarily on the study of black holes. When astrophysics discovers all the potential behind this divine architecture, we will certainly have a precise answer to the theory I am proposing in this treatise. Perhaps, on the other hand, when a star implodes and forms a black hole, this is also explained by the perspective of the gravity of this predecessor universe that exerts a strong influence on the formation of this strange phenomenon, which we can imagine as openings to the dead womb of the universe.

Traditionally, the big bang theory has been the backbone of cosmology, providing us with a model of a universe born from a singularity, which has been expanding for approximately 13.5 billion years. However, in light of new evidence, it becomes increasingly clear that this narrative faces significant challenges, making room for a new perspective: the theory of the "dead universe" that I propose.

The dead universe theory suggests a radically different approach. Instead of conceiving the universe as the result of an explosion, it proposes that the universe is a vast and possibly eternal continuum, where concepts of beginning and end are relativized. This is not just a vague hypothesis; the discoveries of the James Webb offer concrete evidence that challenges the fundamental premise of the big bang. Ancient galaxies that should exhibit signs of interactions and mergers, as predicted by the standard model, remain surprisingly intact, suggesting a much more complex and less linear cosmic history.

The observation of astronomical objects that appear to be older than the age of the universe defined by the big bang model poses a significant challenge for contemporary cosmology. How can the existence of these mature structures be reconciled with a universe that, according to current estimates, is approximately 13.5 billion years old? The hypothesis of the "dead universe" seeks to address this contradiction by proposing that such galaxies are not simple discrepancies, but rather indications of an ancestral universe, whose timeline extends beyond the temporal scale demarcated by the event of the big bang. This theory suggests that conventional cosmological chronologies may need to be revised in light of new evidence, possibly expanding our understanding of the history and evolution of the cosmos.

Moreover, the supposed uniform expansion of the universe, a cornerstone of the big bang model, is called into question by recent observations. Distant and ancient galaxies do not behave in a way that would corroborate a constant and accelerated expansion. This raises a fundamental question: what if the universe is not expanding uniformly, or even if it's not expanding at all? My theory suggests that the cosmos may be in a more complex and static or inverse state than previously imagined, a state where time and space are not absolute, but relative and interconnected in ways that we are still beginning to understand.

This is not just a challenge to the dominant narrative; it is an invitation to radically rethink our understanding of the cosmos. The theory of the "dead universe" offers a path to explore these questions, proposing a "timeless" universe or one that generates its own strange body, light, as its primordial nature was not light, but rather the darkness of dark matter and supermassive bodies, where the beginning and end are human concepts, not universal realities.

From the perspective of the dead universe, the fusion of black holes and the consequent creation of stars may be considered incomprehensible events for beings inhabiting this universe. Imagine a civilization evolving amidst eternal darkness, where light is an abstract, almost mythical notion. For them, the sudden emergence of luminous points in the sky would be beyond comprehension, an anomaly in a predominantly dark environment. Perhaps the equation of UNOl matter also resembles a window tint or solar control film, so that when we are inside we perceive the existence of light, but if we look from the outside in we perceive no light at all, and everything seems to us without light and in darkness. Therefore, a universe immersed in death immersed in a dark fabric may present a reality of splendid light that we cannot see because of the presence of a matter that I describe as UNO.

The theory of the dead universe proposes a new interpretation of the observational boundaries of the universe through an analogy with window tint film. We argue that dark matter and other cosmic anomalies may be analogous to layers that, although transparent from the inside, are opaque when viewed from the outside. We explore how this metaphor can be applied to the study of astrophysics and offer insights into the properties and behavior of dark matter. Just as an internal observer perceives light through a layer of window tint, while from the outside transparency is obscured, our visibility of the cosmos may be limited by material layers that are not immediately apparent to our conventional detection methods.

The theory of the "dead universe" suggests that we live in a remnant of a previous cosmic reality, where dark matter acts as a "cosmic window tint" that distorts our perception of the universe. This matter not only influences the trajectory and speed of galaxies but may also be the reason why we observe the universe in such a dark and enigmatic manner. Gravitational waves and other observations can be seen as the light that permeates this dark layer, offering glimpses of the underlying structure of the universe. Our understanding of the universe's expansion and the distribution of dark matter may be enriched by considering the idea that, just as light passing through window tint, there is an inherent luminosity and active phenomena beyond our current vision awaiting discovery. Therefore, future research should focus on penetrating this "cosmic window tint" layer, revealing the true extent and nature of the universe in which we reside.

Cosmological theories that propose various forms of "barriers" or transition zones in the universe. For example, the event horizon of a black hole acts as a point of no return where gravitational attraction is so strong that not even light can escape, making it invisible from the outside. This is similar to looking at a dark window from the outside; you cannot see, suggesting an absence of light or activity when, in reality, there is hidden wealth.

Extending this to your notion, if there were a "UNOl matter" that acted as this kind of cosmic hue, it could be something that exists within the structure of the universe—a hypothetical substance or field that interacts with light and other forms of energy in a way that masks the activity or underlying structure of the cosmos when viewed from a certain perspective. Such material could theoretically be responsible for the phenomena we observe, such as the effects attributed to dark matter, which influences the movement of galaxies and yet emits no detectable light or radiation, remaining "UNO" or "invisible" to our current obser-vation methods.

The notion of a "domain wall" in cosmology is a hypothetical structure that could act as a boundary between different phases or types of vacuum states in the universe, similar to the interface between two bubbles. It is a speculative concept but one that could potentially explain cosmic separation or transition areas, much like your concept of "UNOl matter" film. Note that while analogies can be useful for illustrating concepts, in scientific publications, they are typically used sparingly and always anchored in rigorous argumentation and empirical evidence.

Furthermore, the very absence of light as a primordial element may challenge the fundamental laws of this dead universe. While they inhabit a domain where darkness reigns supreme, the presence of light could be seen as an intrusion or even as a metaphysical impossibility.

These reflections lead us to question whether we can truly com-prehend the entirety of the universe from our limited perspective as observers of the cosmos. What we consider universal truths may be just a small fraction of cosmic reality, and the dead universe may represent a spectrum of existence that escapes our full understanding.

Perhaps the very nature of light is indeed opposed to the essence of the dead universe. The mergers of supermassive bodies and black holes, which were the original nature of this universe, generated light, an object strange to its reality. This universe will persist forever, immersed in its own eternal darkness, while light shines in contradiction. However, this does not mean that our observable universe is the essence of this dead universe. The mergers and anomalous behaviors of particles altered the original order of this universe, giving rise to strange bodies, such as the galaxies we observe. In this sense, we are mere intruders of chance in this reality, unless there is a creator entity for the dead universe.

Light is something strange to the reality of the dead universe, if we may say so, as it will always exist with its own nature and laws, and it is calling this strange universe that has light as a primordial factor to its nature and essence. In this sense, it is not up for discussion the existence of humanity and life as we know it. "No one can deny that the universe is more for darkness, chaos, and obscure mystery than for a reality of light, as expressed by the wise Solomon. 'The Lord said that he would dwell in darkness.'" (1 Kgs 8:12).

It is an exciting time to question, explore, and perhaps discover the true nature of the cosmos. Our time will always be the present because we are within the eternal time of the dead universe.

Physics deals with the enigma of dark matter. It is conjectured that such matter may consist of compact and supermassive objects, such as primordial black holes, or perhaps hypothetical and indescribable particles known as sterile neutrinos. However, the very concept believed to elucidate dark matter finds a stronger resonance within the scope of the dead universe theory than within the limits of the big bang paradigm. The existence of a past and extinct universe, devoid of all luminance, supports the belief that this process generated energy, similar to the unexplained cosmic puzzle of dark energy. According to this theory, dark energy is not the agent of universal expansion but rather the residual laws of the preceding universe in force.

If we assume that the predecessor universe contributes an additional gravitational effect that alters the apparent rate of expansion, I can add a term to the Hubble Law to represent this. Consider the following hypothetical equation, which includes such a term:

$$v = H_0 \times d + FDU(d)$$

**Where:**

v is the recessive velocity of galaxies, **Ho** is the Hubble constant, d is the distance to galaxies, **FDU(d)** is a function I define to model the influence of the dead universe on the recessive velocity, which may depend on the distance. I define the function **FDU(d)** as follows:

$$FDU(d) = \alpha \times (MDU / d^2) + \beta \times SDU$$

**In this case:**

$\alpha$ and $\beta$ are constants to be determined empirically, **MDU** represents the combined mass or gravitational influence of the dead universe, $1/d^2$ represents the inverse square law of gravity, assuming that the influence decreases with the square of the distance, **SDU** represents structural or quantum effects of the dead universe that may affect the recession.

In the proposed equation, the term $\alpha \times (MDU/d^2)$ could represent a gravitational effect from the predecessor universe, while $\beta \times$ **SDU** could represent additional unknown structural or quantum effects.

To test this modified law I am proposing, I will need to: Determine appropriate values for $\alpha$, $\beta$, **MDU**, and **SDU** based on observational data.

Make predictions about the recessive velocities of galaxies at various distances that can be verified against astronomical observations.

Adjust the values of $\alpha$, $\beta$, **MDU**, and **SDU** to better fit the observations if necessary.

This theory is elegantly simple, easily testable through scientific experimentation. While empirical evidence remains paramount, the big bang theory has imbued the scientific domain with such complexity that its acceptance often relies more on the intricate explanations it provides for cosmic phenomena than on its fundamental merits. The dead universe theory elucidates the pre-big bang epoch—if we may call it that—and suggests that our cosmos emerged not from an original singularity but from the cataclysmic demise of the ancient universe, avoiding the continuous expansion punctuated by explanatory gaps.

The dead universe theory takes into account dark matter, radio waves, and particle behavior. My conviction that our universe is descended from its predecessor is supported by the biblical account of Gen 1:1–2. However, a creative agency does not nullify the dead universe theory for purely scientific purposes. Science does not strive to substantiate the existence of the divine; it merely seeks to investigate natural phenomena and elucidate them through the lens of empiricism. Likewise, it does not exist to deny the divine. Therefore, let us set aside that which eludes explanation and channel our energies toward what can be explained—toward the dead universe theory.

The theory of immensely magnitude gravity of the predecessor universe can naturally warp space-time, a phenomenon known in astrophysics as a "gravitational well," responsible, for example, for bending light. The idea that the observable universe is within the womb of a dead mother universe that died trillions of years ago, the same fate as our universe, which emerged from the womb of the previous mother, may explain what astrophysics has not been able to. The gravitational force of the ancient universe can bend the fabric of the universe in such a way that it creates a "slippery" advancing through space without actually moving. The big bang theory, while accepted to explain the origin of the universe, has gaps, such as the lack of explanation for continuous expansion.

Studies involving particle accelerators, which evidence phenomena similar to micro-explosions, can be interpreted as support for this alternative hypothesis. If the observable universe emerged from a "dead universe," such an event could be interpreted as an expansion driven by the remaining action of the gravity of a previous universe, a concept that could be inferred from the presence and behavior of black holes, which offer indirect evidence of this process. The continuity of gravitational laws, which seem to govern without alteration since the primordial state, may be a testimony to the deep connection between the current universe and its possible origin in a previous and broader context.

A pertinent question regarding the challenge to the big bang model lies in the observation that expansions resulting from explosive events usually introduce a level of randomness in the movement of involved particles. However, the expansion observed in the universe suggests a more orderly and systematic progression, possibly guided by principles not yet fully elucidated by contemporary physics. Regarding the characterization of the "explosion" associated with the big bang itself, the term may be considered inadequate if interpreted in the light of conventional explosions. If such an event does not fit within the traditional parameters of an explosion, what then are the physical mechanisms supporting such a model? The proposition of the big bang, which postulates the expansion of the spacetime fabric itself, demands a source of energy capable of enabling such a phenomenon.

Furthermore, the process described by the big bang does not correspond to an explosion within a pre-existing space, but rather to the expansion of the spacetime structure itself. In this context, the hypothesis of the "dead universe" offers an alternative explanation that could provide a detailed description of cosmic expansion, filling gaps left by the big bang model, which sometimes seems to oscillate in its explanations about the exact nature of the initial event.

Additionally, the regularity and organized structure we observe in the cosmos may seem antithetical to a chaotic and random origin suggested by a conventional explosion. Scientific studies, including those grounded in principles of quantum physics, have indicated that the nature of the universe may incorporate explosive aspects. Consequently, if the observable universe is influenced by a previous cosmic legacy, then the initial conditions and physical laws of this preceding universe could be the regulating keys of the expansion we witness today.

The theory of the "dead universe" not only challenges the foundations of the big bang but also offers more coherent explanations for the existence of celestial phenomena. By proposing a new model for the origin of the universe, this theory paves the way for a deeper and possibly more accurate understanding of the cosmos, transcending the limitations of current science.

Exploring the traditional theory of the universe's expansion in light of the first verses of Genesis, seeking to understand how scientific theories like the big bang relate to the biblical narrative, should be considered a bilateral position. This approach will lead us to reflect on the intersection between faith and science, highlighting that Christianity, far from being an obstacle to scientific development, has significantly contributed to it. It is important to emphasize that many of the most renowned institutions of higher education in the world were established by Christians with the purpose of developing scientific thought. This work also aims to be a manual of faith rationalized in sacred Scriptures.

Despite observations indicating galaxies drifting apart from each other, I remain skeptical of the expansive model suggested by the big bang. Grounding myself in the theory of the dead universe, I venture, speculatively, the possibility that the universe is actually condensing towards a singular point. The dead universe theory could clarify the existence of billions of supermassive objects, black holes, radio waves, echoes, dark matter, dark energy, and even reveal that a UNO and 'invisible' substance permeates the known cosmos.

Upon contemplating the cosmic microwave background, we observe the primordial vestige of the inaugural luminosity of the cosmos, which permeates the universe uniformly—a true cosmic fossil. This

evidence harmoniously aligns with the dead universe theory I am proposing. It posits that the CMB is the indelible impression of a previous
stage of the universe, an era of quiescence or balance before a significant
cosmic transition. The neutrality of the first atoms metaphorically reflects the universe in a state of latency or "death," a phase preceding the
complexity and structuring observed today. This speculative theory suggests that we live only one phase in a much broader, and perhaps eternal,
cosmic progression, where the universe oscillates between periods of
activity and inactivity.

A testimony of creation ex nihilo, that is, the emergence of the universe from nothing. Nevertheless, its expositions find acceptance among
academic and secular circles, while congruent narratives of theological
nature are often marginalized. However, it is undeniable that ancient
wisdom had already consigned the truth that the cosmos emanated from
nothing, a truth that has persisted for almost four millennia, preserved in
the most sacred Scripture and one of the most venerable tomes in human
history: the Bible.

Arguing that Christianity was an obstacle to the advancement of
science is an ignorant and superficial misconception. Neo-atheism, in
particular, has propagated a dishonest correlation between belief in God
and a lack of interest in science. Christians often face discrimination in
universities founded by their predecessors, while arrogant neo-atheists
tend to associate intelligence with the absence of religious belief. However, it is important to recognize that many of these critics of the Christian faith received their education in institutions founded by Christians,
which contradicts their narrative of intellectual superiority.

Furthermore, many of the scientific advances and theories that
these critics use in their arguments were developed by Christian thinkers
throughout history. Therefore, it is questionable to suggest that belief in
God is an obstacle to scientific thought when science itself owes much to
the contribution of religious scientists.

Investigating the trends of neo-atheism, I came across the renowned
biologist Richard Dawkins and his documentary *The Virus of Faith*.
Dawkins, though respected in his field, presents a one-sided and critical
view of the theological roots of faith, focusing on superficial criticisms
rather than deeper approaches. His work seeks to establish a reductionist link between faith and science, two distinct and complex fields that
should not be compared in such a reductionist manner.

In his documentary, Dawkins portrays faith as a harmful element to scientific progress, a view that I consider uninformed and reductionist. He ignores the richness and depth of the spiritual experience, reducing it to a mere illusion or superstition. Furthermore, by elevating science as the only valid source of knowledge, Dawkins disregards the significant contribution of religion to human understanding and morality.

Catholic monks played a significant role in the development of the first universities in medieval Europe. Many of the earliest universities were founded by religious orders, especially monastic orders, which were centers of learning and intellectual production in the Middle Ages.

Dawkins also resorts to questionable strategies, such as using personal testimonies to support his narrative. By portraying religion as an obstacle to scientific progress, he disregards the fact that many of history's greatest scientists were people of faith. His attempt to polarize faith and reason is a crude idea that is detrimental to constructive dialogue between the parties.

Moreover, Dawkins seems to ignore the complexity and diversity of religious beliefs, reducing them to naive stereotypes. Religion is not monolithic, and its contributions to society go far beyond what Dawkins suggests. By portraying faith as a virus, he disregards the richness and depth of spiritual and moral experience.

Dawkins fails to provide a critical and detailed analysis of the relationship between faith and science. His generalistic approach and polarizing rhetoric hinder constructive dialogue and mutual understanding between believers and non-believers. To advance this debate, it is essential to recognize the complexity and diversity of religious beliefs and to avoid simplified and uninformed generalizations.

The analysis of faith has two perspectives when grounded in the fear of divine punishment proposed by Dawkins, versus faith based on the love and fear of God taught by Christianity; this is crucial for understanding the different perspectives within monotheistic religions. However, it is important to recognize that this analysis must be done carefully, using a variety of linguistic and cultural resources to support our arguments.

Criticism of the "opportunistic aspect of religion" often exposes the weaknesses of the religious character of its leaders, highlighting the mistakes of blind fanaticism and false religiosity. It is essential to understand that not everything proclaimed as a spiritual representation of God is genuine, often being just a manifestation of perverse human character.

When evaluating religious experiences, we must consider the linguistic resources of the books used in each of them and the anthropomorphisms created to express man's relationship with God. Although biblical accounts, such as those of Moses, may be accurate in their description of the divine, it is important to recognize the linguistic and intellectual limitations of the time in which they were written.

Fear and awe, although related in Hebrew vocabulary, can convey distinct feelings in spiritual experiences. For example, Paul, despite being versed in various languages, recognized the difficulty of expressing his spiritual experiences in words.

A faith based on fear is false, as it contradicts the essence of the relationship with God proposed by Christianity, which is grounded in love and fear of Jesus Christ.

Religion, often selfish and manipulative, creates rules and dogmas that distort the true nature of the relationship with God. Jesus, as a mediator between God and man, offers this communion at no cost, without the need for intermediaries or fees.

The crisis of atheism arises when the atheist positions himself as the exclusive holder of reason, underestimating other forms of knowledge and expression. A deeper and more sincere approach to the question of God is essential for truly grounded atheism.

Negative feelings towards religion often stem from distorted interpretations and bitter experiences. Atheists should seek a new, contextualized approach to the contemporary world, avoiding generalizations based on past experiences.

It is crucial to recognize that emotions and feelings are valuable tools for expressing our humanity and our relationship with the divine. Reason should guide these expressions, ensuring they are aligned with truth and authenticity.

Christians do not impose any kind of psychological debt and are not agents of any kind of sick alienation forged by faith. The preconceived accusations of atheists would be countless particular promises made by Jesus Christ to his disciples with the pretense of deceit. Some atheists disdain this absurdity through a fallacy illustrated from the conception of fictitious credit from a financial institution, without a mature and realistic understanding of the theology of salvation by grace through faith in the resurrection of Jesus. Prominent figures like Richard Dawkins and Sam Harris often challenge these views, emphasizing the importance of empirical evidence and rationality in belief formation.

We cannot conceive the madness of creating preconceived definitions of deep concepts of Christian faith just to receive applause from peers. Another irresponsible conclusion of Christian faith studies in a frivolous way is British writer Patrick Condell, who offers a critical view of Christian thought. In a troglodyte investigation, he assumes a superficial analogy of Christian theological thought, further affirming with the utmost cheekiness the possibilities of Christians facing a serious problem with this type of belief in Jesus. In his nihilistic line of reasoning, "a default on the debt paid by Jesus," not to mention even the annoyance of the figurative representation of sprinkled blood, will put the Christian in great trouble with God. In truth, he expresses attacks on theological issues as if he were speaking to small children playing in the village of Nobodys, devoid of the noble loving lucidity of a great spirit in search of peace and without the noblest judgments of reason. He uses the masks of disbelief in God as an argument to justify a desperate impossibility.

It is not a good response to refute the truth by engendering lies, as if Christianity were strengthened by eternal threats of hellfire. Would it be a spectacular trap to structure your life on the teachings of Jesus Christ? Are the pillars continuously presented by the Christian church crude? Will God fry us for all eternity in hell if we deny the message of salvation in Jesus? Richard Dawkins and Daniel Dennett, notable critics of theism, highlight the lack of empirical basis for such claims, arguing that religion often relies on fear and manipulation to perpetuate its beliefs.

According to Patrick Condell, the awaited destiny for those resistant to the word of salvation of Jesus Christ is agony and eternal suffering. And that Jesus will do nothing about it. In our Christian proposition, he has already done everything well. I reaffirm this truth, for he is right. If we consider that Jesus Christ has already done everything for humanity on the cross of Calvary. Condell is fully right to say with his British mentality of bankrupt financial institutions extorting their clients: "We have no way to pay off this debt settled by Jesus." Look at what a wonderful sentence; therefore, we always need the continuous forgiveness of Jesus Christ. The Bible informs us of special reasons about eternal divine forgiveness, and its infinite mercies. Rejecting the person of Jesus Christ would be an option of God, or would it be a choice? Would denial be an act of disbelief in the ability to believe in continuous forgiveness?

The debt of human sin is something unpayable, and when the individual accepts the payment by the sacrifice of Jesus Christ, he becomes free from this guilt. The ancient creditors, in the case of the devil and his fallen angels forgotten by Patrick Condell, according to the sacred Scriptures, can no longer claim payments for sins. Authors like Richard Dawkins and Christopher Hitchens may argue against the notion of original sin, questioning the need for a savior and emphasizing individual responsibility.

Theoretically, there is a debt settled by Jesus, and denying Jesus as the eternal savior is the atheistic way to invalidate forgiveness. We have the testament of sons and evidence in documents of the price paid in blood. This denial of settlement becomes ridiculous. The price paid by Christ was the currency called life.

Critics of Christianity, such as Richard Dawkins and Sam Harris, think they can offer an alternative view of redemption and morality, questioning the validity of religious narratives and defending a secular approach to ethics, a proposal for a rationalized "crucifixion" in values that are embedded in Christianity.

There is no need for a new crucifixion. The illusion peddler, Patrick Condell, tries to deceive the weak through manipulation without understanding the doctrine of Christ. Rational faith is not enslavement but is potentiated in liberation in the form of certainty of new life in the resurrection of Jesus Christ, for he settled the negative credits of eternal death. Critics may challenge the validity of the concept of rational faith, arguing that religion often restricts freedom of thought and promotes irrational beliefs.

Faith is the practical rule of Jesus Christ's works and is the potential for free access to God. It is not a word translated into language of deceit and danger. Faith is certainty within certainty, and there is no other word to describe this conviction in something proved. It is also the vigorous synonym "antirational" in the face of the opposite called denial for those devoid of this proof. Patrick Condell has every right to make his speculations, but not to condemn an experience in the expectation lived by people. He only tries to support himself in the act of observation of his arrogant science. He becomes a mere thoughtless critic of this indescribable experience. If he himself can see two faces in the same story, establishing the concept of spirituality linked to that of religion, in the next attempt with requirements for surrender to know the premise of faith, he sees disbelief.

In summary, we can bring evidence of self-knowledge through spiritual experiences, and human errors in the way they conceptualize spiritual experiences. In Christian theology, we do not conceive a concept of self-indulgence as a positive phenomenon for extracting our faith. The Scriptures are categorical in affirming that faith is the secure proposal of redemption and perfection of men. It is a Gift from God, and not a formula manifested by collective obsession. It is not an internal representation externalized in the experimental spiritual phenomena of the collective. On the contrary, it encompasses itself in particular expressions, then establishes itself with the practical actions of works and takes real proportions in the concentrations of groups that live the same faith. Accusations of compulsions to move spirituality through faith, in this case, is a cowardly act of self-disbelief. Authors like Richard Dawkins and Daniel Dennett can offer a critical analysis of the religious phenomenon, highlighting the limitations of faith without intelligence, and questioning its validity as a means of knowledge.

Post-modern man has veered into nihilism; he consumes himself in the void of nothingness, swallowing himself in the solitude of existential emptiness. Just as we did not know the existence of black holes and many other cosmic possibilities in the past, we cannot dismiss the possibility of life on other planets just for lack of current evidence. Richard Dawkins argues that disbelief in God is not an absolute denial, but rather a position based on the absence of convincing evidence.

I see that the growing denial of faith in God can be interpreted as a quest for liberation from what some consider a frivolous deception amidst personal frustration. Considering faith as based only on innocence is to reduce its complexity and strip it of its legitimate value. Conversely, by recognizing that faith can also be an expression of human guilt, without the expiation offered by Christianity, we can better understand the nature of this guilt.

Dualism, conceived as the antithesis of the supreme good, has been the object of reflection for many philosophers throughout history. Understanding life and death as intrinsic opposites is a profound philosophical challenge. For example, if we assert that fire is good, that seems evident, but considering the consequences of putting our hands in the fire, we realize that it can also be bad. Thus arises the question: is fire inherently bad or good?

In this context, Christianity sees life as a negation of eternal death, offering a choice of freedom in exchange for something considered superior. The act of creating hell for the destruction of wicked men becomes an act of God's kindness, since the Scriptures affirm that they will go through the second death and be forgotten forever. A just and good God cannot imply eternal suffering for a finite life of sins, for that contradicts the attribute of God's love. The greatest hell for a human being has always been knowing that he has lost what he could gain as an inestimable value; perhaps for some time he will have this perception of the hell of pain and suffering.

Christianity preaches the appreciation of life and the hope of eternity in the resurrection of Jesus Christ. Christianity does not give a gift box with a rabbit and a poisonous rattlesnake inside, for they cannot coexist. But it only warns about the possibilities of a rattlesnake entering its box and striking the rabbit that decided on its own to live inside the box. An alternative view on life and death, I emphasize the importance of appreciating life in this world, regardless of religious beliefs about life after death.

Conducting our lives irresponsibly would be to deny everything we have learned in the sciences over the years. Atheism sins and does not sin, but it makes a grotesque mistake. It doesn't even know where it came from, but it thinks it knows where it's going. The big problem with this question is the certainty of science only by visible experience; it cannot determine anything about God. Any assertion that God is this chance we know can be dangerous; my assertion is that there is no chance where things happen. The greatest coincidence of the universe would be your reading of this text at this exact moment, and thinking it is a coincidence. Taking into account the possibility that you are not a real person.

Authors like Richard Dawkins and Sam Harris can offer a critical analysis of empiricism and scientific knowledge, highlighting the importance of rational investigation and the scientific method in understanding the natural world. However, they have no knowledge of spiritual experiences, and they have the right to attack them, but that does not mean they are right. These men have the false premise of asserting that faith is dying and disbelief is gaining ground in the world. Perhaps they have to go back to basic education to learn mathematics, as proportionally belief in God has increased significantly compared to our ancestors. Today, the world has more than eight billion people, and considering Muslims, Christians, Jews, and other beliefs professing faith in God, perhaps we have more than six billion people who believe in God today, whereas before only a few million believed. There are more people who believe in God today, and the numbers of faith are increasing. Science in no way diminished the number of believers in God; this idea is easily refuted by data.

The data shows a trend contrary to the assertion of neo-atheists that science kills belief in God and religion. On the contrary, scientific advances often awaken greater admiration for the complexity and beauty of the universe, leading people to reflect on spiritual and metaphysical issues. Furthermore, faith and science are not mutually exclusive; many eminent scientists hold deep beliefs in God. It is important to recognize that the pursuit of scientific knowledge and the practice of religion can coexist harmoniously, each addressing different aspects of human experience.

Therefore, the idea that science is eradicating belief in God is deceitful and does not reflect the complexity of the interactions between science, faith, and society. Instead of seeing science as a threat to religion and vice versa, we can see it as a complementary tool that helps us better understand the natural world, while religion offers meaning and purpose to our lives.

Christianity is the great precursor of modern science, and this tradition continues to this day, showing that the relationship between faith and knowledge goes beyond mere speculation. Christian universities, throughout the centuries, have been centers of intellectual and scientific production, where the pursuit of truth is valued in both secular and theological areas. Thus, when considering the impact of religion on academia, it is crucial to recognize the dynamic interaction between religious beliefs and the development of human knowledge.

When considering the most prominent universities in the world, it is evident that a large proportion of them were founded by Christians, both Catholics and Protestants. This directly challenges the idea that religious faith is incompatible with critical thinking and the pursuit of knowledge. Instead, it suggests that faith and reason can coexist and even complement each other in the process of scientific and academic investigation.

However, some individuals, even within the academic environment, perpetuate a dishonest narrative, suggesting that belief in God and the pursuit of scientific knowledge are incompatible. This narrow and prejudiced view, promoted especially by neo-atheism, ignores both the rich Christian intellectual tradition and the fundamental role of Christian universities in shaping brilliant minds throughout history.

It is ironic to observe that many critics of Christianity receive their education in institutions of higher learning founded by Christians, taking advantage of the resources and infrastructure established by those who share the faith they disdain. Furthermore, many of the theories and concepts that these critics use in their arguments were developed by Christian thinkers over the centuries.

Therefore, it is important to challenge the simplistic narrative that belief in God is synonymous with ignorance and lack of intelligence. The history of universities and scientific progress reveals an intricate relationship between faith and reason, where many of the greatest scientific advances were made by deeply religious individuals. By recognizing the contribution of Christians to science and academia, we can open space for a richer and more respectful dialogue between faith and reason, promoting a broader and integrated understanding of the world around us.

Misunderstanding is common among human beings, especially when we observe the disasters caused by followers of Jesus Christ. The actions of some can tarnish the credibility of the Christian faith, such as the cases of pedophilia that shook the Catholic Church. Although doctrines are correctable, emotional scars remain in people.

We need to be cautious and recognize the dangers we face as Christians; there is a strong academic tendency towards discrimination against those who believe in biblical creationism. We do not know the reason, but many Christian communities persist in following misleading teachings. Prosperity theology, for example, has corrupted many Evangelicals, leading them to greed. However, these errors do not invalidate the resurrection of Jesus Christ, which is the foundation of the Christian faith.

The errors of Christianity should not serve as justification for atheism. Many atheists base their disbelief on the mistakes committed by religions, especially Christianity. However, this is not a valid reason to deny the existence of God. If atheism is based on disbelief in God due to human errors, then it is an emotional response to negative events.

The search for evidence of God's existence can be challenging, but we must consider the testimony of Jesus Christ resurrected as the greatest evidence. Ignoring this evidence means ignoring one of the fundamental bases of the Christian faith.

It is important to understand that the understanding of God goes beyond human capabilities. He manifested himself in Jesus Christ, but many still seek God elsewhere. We must look at the *fact* of the resurrection as the main evidence of the existence of God.

Ultimately, does God exist? If we accept the testimony of Jesus Christ as true, then God exists. We must base our understanding of God on the *fact* of the resurrection, rather than relying solely on human reason.

Why believe in the existence of God? And if we believe, what path should we follow to deepen this belief safely? What reasons do we find in our rational search to guide true spirituality? And, finally, do we have evidence grounded in facts confirmed by historical witnesses to sustain our faith in the chosen belief?

A distinctive landmark of Christianity compared to other religions is its long history of worship of the One God, witnessed by thousands of real people. The historical truth of this worship is not based solely on the experience of a single individual, like many other religions, but rather on the history of a people. From Enos, considered the initiator of the worship of God Yahweh, to the time of Jesus Christ, the history of Christianity is deeply rooted in the collective experience of a community of faith.

On the other hand, Islam, the second largest monotheistic religion after Christianity, has its theological bases partly inspired by the history and theology of biblical authors. However, its origin dates back to the seventh century in the Arabian Peninsula, with Muhammad receiving a divine revelation in isolation on Mount Hira. Unlike the eyewitnesses of Christian history, the Quran was revealed to Muhammad through an archangel in secret, without direct witnesses.

The decision to choose in whom to believe in the One God does not offer many options: Jesus Christ resurrected or Muhammad? While Muhammad received the revelation of the Quran in isolation, without eyewitnesses, millions of people witnessed the life and resurrection of Jesus Christ. More than fourteen thousand ancient texts, in addition to the testimonies of the apostles and thousands of saints, attest to the resurrection of Jesus Christ.

Therefore, when considering in whom to believe, it is essential to study the reasons that will underpin this faith. We must seek concrete and historical evidence, rather than relying solely on personal experiences or religious theories elaborated from isolated revelations. The resurrection of Jesus Christ is the only solid evidence that sustains the Christian faith, providing a reliable basis for those seeking evidence of God's manifestation.

We have reached a point in our lives where we make decisions about which paths to follow, based on what we consider to be correct. However, the question arises: what is the evidence that corroborates this truth? Can we trust it simply because it is a tradition passed down by our ancestors? Or does it deserve our trust because it seems to provide satisfactory results? Religions, whether right or wrong, strengthen themselves according to our choices, and we are responsible for the growth of religions. Who really cares about the truth? The early Christians, who brought the message of Jesus to the pagans, understood that he was the truth and were willing to die for that redeeming message of the cross.

People are always in search of better alternatives, often without considering if there is a central truth. We live in a culture that values freedom of religious choice, where new options constantly arise for all tastes. However, few care about the truth itself, resulting in a multiplicity of religious options. For many, what matters is simply filling an inner void, a selfish sensation that reflects the values of capitalism.

Christianity, in part, succumbed to this trap, setting aside truth in pursuit of satisfaction. In a scenario where the search for better options to meet needs has become the dominant game, most make their choices based on others' opinions, without caring about the question of truth. When Jesus proclaimed to be "the way, the truth, and the life," he was highlighting the path of sacrifice. Thomas, among the disciples, was someone with solid convictions about the truth. We should thank God for Thomas, an apostle who sought concrete evidence of Christ's resurrection. While the other disciples maintained their unwavering faith in Jesus, without questioning, Thomas sought a deeper and more enlightening understanding of the truth. He longed to see and touch the resurrected Jesus, concerned about the religious fervor of the other apostles. We can criticize him for his disbelief, but his quest for concrete evidence should inspire us to cultivate a deeper faith in God. However, we can also recognize that his doubt was cleared when Jesus appeared to him, showing that he was wrong about the resurrection of Christ. Thomas had a science-oriented and experimental mindset. No one could convince him with just reports or testimonies about the resurrection; he needed tangible evidence to be sure of the truth of the fact. It was necessary for him to touch the body of Christ to confirm that it was not a collective illusion or a spiritual manifestation. By feeling the physical presence of the resurrected Jesus before him, with his hands touching the flesh and bones, Thomas had unwavering certainty that Christ was truly alive. This transformative experience not only dispelled his doubts but also strengthened his faith and conviction in the resurrection of Jesus.

The resurrection of Jesus Christ is the greatest proof of the existence of God and the greatest difference between Christianity and other religions. The apostles preached this central message. Thomas was the first to be certain of Jesus' resurrection. He needed concrete evidence, and when he touched Jesus, he understood that he was real.

If Jesus is not the greatest evidence of God's manifestation, there is no other way to reach God. He is the historical Fact, the divine evidence presented to men. Although other religions may contain relative truths, the path to a new life begins with Christ. He is not just an option but the way and the absolute truth. The other truths are only shadows of his representation.

Christianity preaches Jesus Christ resurrected from the dead, the only one who gave his life for the cause of eternal life. He is the evidence of truth, and we, like his followers, express this truth by giving our lives for it. This is our way of expressing what is true, the evidence of the Fact.

As we examine Gen 1:1–3, written by Moses, we are confronted with the theological richness of the creation of the cosmos. It is important to consider both scientific theories and the biblical narrative when discussing the origin of the universe, following the example of Thomas, who sought tangible evidence to support his faith. We cannot simply accept blindly that God created everything without questioning. On the contrary, it is essential to explore the nuances and complexities of both science and theology to obtain a more complete and profound understanding of creation and the Creator.

As we explore the concept of the big bang, we will examine how this theory may be aligned or distant from the biblical perspective on creation. Our goal is to critically analyze how scientific discoveries can enrich our understanding of God's creative work and promote constructive dialogue between science and theology. By addressing the preexistence of Christ and his relationship with creation, we recognize the complexity and beauty of the biblical narrative and seek to find points of convergence between science and faith. It is essential to respect different perspectives and approaches when dealing with such profound questions as the origin of the universe. By uniting scientific knowledge and faith, we can enrich our understanding of the world around us and of God's nature as the supreme Creator.

Sometimes, people face difficulties when studying biblical texts. It's interesting to note how the Scriptures are organized and why Moses structured them in such a poetic sequence, starting with the creation of light, or perhaps it was the later scribes who arranged everything. Understanding the organization of the texts is crucial for grasping the message of creation. For example, it's something I intend to delve into further so that you can better understand. Personally, I revisit the chapters of Genesis about eight hundred times a year. I do this with the aim of understanding and internalizing the truths contained therein. It's part of my daily routine, both in the morning and in the afternoon and evening, to seek improvement and inspiration, often returning to these verses even as I'm reading other books, in the hope of synthesizing truths and receiving guidance guided by the Holy Spirit.

This is a subject I have been teaching for a long time, and the word of God is infallible and profound. Each time you study, you gain a different perspective on the text and what God is trying to communicate, administer, and teach. In the contemporary world we live in, studying about creation is crucial. People in the church often face criticism regarding creationism, but those who criticize usually do not present proven hypotheses, while others use only philosophical vanities, trying to discredit biblical creationism only with the strength of argument.

Quantum mechanics, for example, tends more to add to the faith of Christians than to distance them from biblical creationism, as much as physics and cosmology. The problem lies in interpreting the biblical text to support dogmas, which can pose a greater danger to Christians than scientific theories that are not being developed to conspire against religious doctrines.

I have had a great interest in quantum mechanics, physics, and cosmology since I understood the universe, as they complement and enhance my understanding of creationism. In this way, I can easily allow the coexistence of a rational faith with that generated by hearing the word of God.

One of the things that made me happiest, especially in the nineties, was when they were talking about creating the LHC, the Large Hadron Collider, where they would, through this accelerator, generate a collision between particles to simulate the creation of the cosmos and the universe. I thought, "These guys don't know what they're getting into." And today, the results we have in modern physics increasingly point to creation. That's incredible! I remember 2008, when a picture of the particle accelerator came out and became the wallpaper on my computer. I believe it's the eighth wonder of the world and would love to visit it one day.

The LHC, known as the largest global scientific endeavor, achieved an unprecedented feat by conducting the first controlled supercollisions of particles in a laboratory environment. These collisions, performed in the colossal particle accelerator, recreated the extreme conditions that prevailed at the instant of the big bang, which occurred approximately 13.7 billion years ago. Located in a monumental underground tunnel, 27 kilometers long, the LHC achieved an unprecedented energy of 7 trillion electron volts, resulting in the generation of "scaled-down big bangs." The data generated by these collisions will be meticulously analyzed by thousands of scientists in the coming years, representing a significant advance in understanding the origin of the universe and matter. This achievement marks a new stage in the exploration of modern physics.

I understand that the study of creation and the relationship of Christianity with scientific theories is really important and often underestimated. The geocentric theory, for example, was a view adopted by the Catholic Church at a certain moment in history, but which nowadays is considered outdated in light of scientific evidence. It is important to recognize that science and theology can offer complementary perspectives on the universe and creation, and that both have much to contribute to our understanding.

The theory of geocentrism, based on biblical passages, such as the one found in the book of Joshua, lacks scientific foundation and must be understood within the historical and literary context of the text. The book of Joshua is a historical account of the victories of God's people and does not intend to offer a scientific description of the structure of the universe.

It is important to emphasize that the geocentric theory was not developed by a Christian, but by the Greek astronomer Claudius Ptolemy, who based himself on the theories of Aristotle and Plato. The Catholic Church adopted this theory as dogma, but this does not mean that it originated from Christian sources. Furthermore, the rule of biblical interpretation that requires clarifications of obscure texts through other texts does not find support for the defense of geocentrism in Scripture.

On the other hand, the heliocentric theory, which postulates that the Earth and the planets revolve around the Sun, was founded by Nicolaus Copernicus, a devout Christian and scholar. His contribution to astronomy challenged the prevailing geocentric conceptions of the time and paved the way for a more accurate understanding of the solar system.

In addition to Copernicus, other Christians such as Giordano Bruno also made important contributions to the development of astronomy, even facing challenges and criticism from the church and society of the time. Their research and ideas, although often misunderstood, contributed to the advancement of human knowledge about the cosmos.

The development of scientific knowledge often involves overcoming outdated conceptions and seeking new perspectives on the universe. An example of this is the case of Claudius Ptolemy, who, despite formulating a mistaken theory, was initiating a process of reflection and questioning about the nature of the cosmos. This attitude of questioning and seeking understanding is essential for the advancement of science.

However, it is important to recognize that science has its own permissibility for error. Often, scientific tests for medications or research on diseases can be mistaken, leading to inadequate or even harmful results. A notable example is the case of the drug Vioxx, which was withdrawn from the market due to findings that it increased the risk of heart attacks. This error in science had serious consequences for public health and underscores the need for constant vigilance and review of scientific methods and results.

On the other hand, it is curious to note that while errors in science may be accepted as part of the learning process, the same is not so tolerated when it comes to issues related to Christianity. For example, the geocentric theory, which was adopted by the Catholic Church as dogma for centuries, was later refuted by the heliocentric science of Nicolaus Copernicus. Although this error has been corrected over time, it is often used to discredit Christianity as a whole, while errors in science are seen as part of scientific progress.

This discrepancy in the perception of error between science and religion highlights the importance of an open and respectful dialogue between both fields. Both areas have much to contribute to the understanding of the world and human nature, and it is essential to recognize that neither is immune to errors or misunderstandings. What matters is the continuous pursuit of truth and knowledge, regardless of personal beliefs or historical context.

However, it is regrettable that Christians often face discrimination and prejudice in universities, where the idea that a Christian cannot be a scientist is prevalent. This misguided notion creates a hostile environment for those who wish to reconcile their faith with a scientific approach to the world. The bullying and mockery faced by Christians in these institutions are reflections of this narrow and intolerant mindset.

It is important to note that the big bang theory, initially proposed by a Christian, is not intended to oppose creationism but rather to offer a scientific explanation for the origin of the universe. Georges Lemaître, a Catholic priest, developed this theory, which was praised by Albert Einstein as a beautiful explanation of creation. However, despite this significant contribution by a Christian to science, stigmas and prejudices against Christians in academia still persist.

Whenever a devout Christian, a man of God, inspired by the Holy Spirit, creates a theory, this is rarely mentioned. Being a believer is often associated with being crazy, irrational, incapable of thinking. A common injustice is to ignore Father George Henry Lemaitre, who was the theorist behind the idea of the expansion of the universe, also known as the big bang theory. He published a scientific paper on this theory long before being recognized by an American who later popularized the idea. Lemaitre created this theory to show the power and glory of God in creation, but his contribution is often ignored or minimized.

These examples highlight the importance of not unfairly judging or criticizing those who have religious faith. Often, their beliefs can inspire significant scientific discoveries and contributions to the advancement of human knowledge. It is a reminder to keep an open mind and recognize that truth can be found in various forms of thought and perspectives, including religious ones.

It is essential to promote an environment of respect and tolerance in universities, where all individuals, regardless of their faith or belief, can feel encouraged to explore the world around them with an open and curious mind. Dialogue and mutual understanding between science and religion can enrich our understanding of the universe and our own existence.

Thus, it is crucial to recognize the role of Christians in the pursuit of scientific understanding of the universe, as well as to avoid simplistic generalizations that disregard the historical and philosophical nuances involved in the debate between faith and science. It is essential for the church to engage more in the study of creation, seeking to understand and dialogue in an informed manner with scientific advances. This will not only enrich our understanding of the world around us but also help us build bridges between faith and science, promoting constructive and enlightening dialogue.

"In the beginning, God created the heavens and the earth. The earth was without form and void, and darkness was over the face of the deep. And the Spirit of God was hovering over the face of the waters. And God said, 'Let there be light,' and there was light" (Gen 1:1–3). Understanding the profound meaning of these verses may require not only a deep study of quantum mechanics and physics but also a spiritual connection to the Holy Spirit for a fuller interpretation.

For God is communicating many things in this text, in this verse, and in other parts of Genesis, chapters 1 and 2, and they may go unnoticed. Just as many theories of quantum mechanics and physics, of advanced sciences, I, as a theologian, would not understand unless they were explained. So they will not understand in depth, but the biblical text is very rich, especially when studied from the Hebrew, to understand the verbs being employed in the text, which will bring more light and understanding.

Today, if you pay close attention, this interpretation of the Holy Spirit will cause you to grow in grace and knowledge. The first thing the Scriptures say, in Hebrews, chapter 11, verse 1, is that by faith we understand that God created the world, the visible things from the invisible, from what does not exist, from the Latin *ex nihilo*. God brings into existence the things that exist from what does not exist.

This is in the Bible, the text in Hebrew is saying this. God brings into existence the things that exist from what does not exist. It is a fundamental perspective. Imagine, for example, a quantum mechanic explaining something similar. However, for many believers who do not deeply understand the Bible, they may consider this idea insane. But how can someone be considered crazy when they are a creature of God, illuminated by the Creator himself to discover and understand the world around them? The Earth, the universe, the entire cosmos, are available to be studied and examined, thus revealing the majesty and wisdom of their Creator.

God has given capacity to his creation. We are all endowed with abilities and capabilities, regardless of our faith. However, for the Christian, there is a significant difference: they are also empowered by the word and the breath of life, which is the Spirit of God.

Let's explore this a little more. A nuclear physicist presents an intriguing question: how is it possible that an empty atom forms the ground that surrounds us? For a long time, it was not known that the atom was largely empty. It was only with the advancement of science, especially in the study of physics and quantum mechanics, that this discovery was made. Projects like the LHC (Large Hadron Collider) were developed with the aim of exploring these questions and seeking answers. One of these inquiries led to the discovery of the Higgs particle, fundamental to our understanding of the origin of all things.

The exploration of the "visible universe" is essential for deepening our understanding of the mysteries of the creation of the sovereign God.

The term "visible universe" refers to the part of the cosmos that we can observe directly, either through terrestrial or space telescopes. This region encompasses everything we can detect through light and other forms of electromagnetic radiation. The visible universe includes stars, galaxies, nebulae, and other celestial bodies that emit light or are illuminated by light sources. However, this observable portion of the cosmos represents only a small fraction of the total universe, as there are vast regions inaccessible to direct observation, due to distance, darkness, or other factors.

Many thinkers may seem eccentric, but in reality, they are revealing a reality that the Scriptures have proclaimed for a long time. The truth is there, just waiting to be discovered through careful study and analysis. For example, the physicist who founded quantum physics stated that there is no such thing as matter. This idea resonates with the biblical concept that God created from nothing, as mentioned in Hebrews chapter 1, verse 1. When Moses uses the verb *bara* to describe creation in Genesis, he is essentially saying that God brought into existence everything that exists from nothing. This confirms the understanding of quantum physics that matter is only a manifestation of forces and energies in interaction, and not something substantial in itself.

According to various theories, including quantum physics, we realize that matter, including our own bodies, is essentially "nothing." This raises profound questions about the nature of creation and the absolute sovereignty of God. If God created from nothing, he must be considered absolute, unable to be identified with nothing.

Furthermore, quantum mechanics leads us to reflect on singularities, such as black holes. Imagine Jupiter, one of the largest planets in the solar system, multiplied millions of times and compressed into a single point, forming a black hole. In this state, gravity is so intense that the planet is reduced to a singularity, an infinitely small point where the known laws of physics seem to fail. This makes us realize that nothing is as we conceive it. But, at the same time, it reminds us that God cannot be reduced to nothing, for he is the Creator and sustainer of all things.

Father Lemaître, known as the creator of the big bang theory, was aware of the importance of grounding his theory, especially considering his faith as a creationist. He proposed that the cosmos emerged from a vacuum, emphasizing that God cannot be associated with emptiness, for he fills all things. This leads us to reflect on the infinite nature of God, who reigns over all dimensions of time and space, as mentioned in the Scriptures.

When I introduced the theme of time, I highlighted that both time and space are creations of God. Quantum mechanics teaches us that even time and space can be reduced to "nothing" when confronted with certain phenomena, such as black holes. This leads us to understand that time is a divine creation, as described in the big bang theory.

The concept of the multiverse, or the idea of universes upon universes, raises intriguing questions about the nature of reality. Imagine yourself in different realities simultaneously: in one you are having coffee, in another you are working on a computer, and in another you are playing soccer. This makes us reflect on the vastness of the universe and the possibility of the existence of multiple realities beyond our comprehension.

Before the existence of time and space, God already existed. He dwells beyond the heavens of heavens, transcending all created dimensions. This view is synthesized in Reformed theology, which emphasizes the greatness and glory of God, which cannot be contained by the cosmos.

The theory of black holes, in turn, continues to challenge our understanding. Physical equations have not yet provided a definitive explanation for this phenomenon, leading some people to speculate about the possibility of them being portals for extraterrestrial visitors. However, this complexity reminds us that the nature of God transcends our capacity for understanding, and that his greatness is beyond the limitations of human science and technology.

During his final moments, reflecting on Albert Einstein's deterministic view, he concluded that perhaps God does not play dice with the universe. His thinking, along with that of others like the Arminians, believed that everything operates by choice and that events could be predicted through mathematical calculations. However, this deterministic perspective contrasts with Heisenberg's uncertainty principle, which states that the behavior of particles cannot be predicted. Similarly, in the spiritual realm, we cannot foresee the events to which Jesus referred in his teachings, such as when he said, "But about that day or hour no one knows, not even the angels in heaven, nor the Son, but only the Father" (Matt 24:36). The uncertainties regarding the destiny of life and the cosmos should serve as a factor for the growth of our trust in God on earth. This theological uncertainty challenges us to exercise faith and trust in God, recognizing the limitations of our knowledge. Through this uncertainty, we gain a deeper understanding of God and our relationship with him.

God dwells above time and space, not confined by them. Scriptures speak to us about creation, highlighting that the universe was brought into being by the word of God. The passage from Heb 11:3 reminds us that faith enables us to understand that the universe was formed from what is unseen. Revelation 4:11 and 2 Pet 3:4 also emphasize God's role as the creator of all things from the beginning.

These biblical texts underpin the idea that God is the supreme creator. However, at times, we encounter passages that leave us perplexed, such as the one mentioning that God created out of nothing. This question has always intrigued me since my conversion. What is God trying to convey to us with this?

Therefore, it is important to remember that God reigns over all creation, including time and space. As Solomon stated, God not only dwells in the heavens but in the heavens of heavens. The mention of the multiverse theory, proposing the existence of parallel universes, adds another fascinating dimension to this discussion of creation and divine sovereignty.

It's important to note that many scientists avoid references to God in their studies, opting for more neutral terms. However, even without explicitly mentioning a deity, their discoveries often align with the biblical teachings on the origin and nature of the universe. The understanding that God is the overarching mind behind all this energy and vibration underscores the biblical view that creation is sustained by a higher intelligence. Therefore, when we look at others, we can recognize the presence of Christ and the Spirit of God in them, as we are made in the image and likeness of the Creator. This advancement in quantum physics, cosmology, and quantum mechanics is increasingly converging with the narratives of biblical creationism. This reality cannot be ignored. And in light of this, new theories and interpretations are being proposed to explain the cosmos and its origin.

Isaiah 40:17 says: "All the nations are as nothing before Him, they are regarded by Him as less than nothing and meaningless."

The biblical text that mentions that all nations are considered as nothing before God has always caused me some frustration since my conversion. I questioned what God meant by this. He regards them as less than nothing, as meaningless.

The concept of "less than nothing" is intriguing, and the Bible presents it in various verses, such as Isa 40:17 and 41:24. These passages highlight the greatness and power of God over all things, including emptiness and nothingness.

Creation reveals God's glory in a silent yet powerful manner. Every aspect of the cosmos bears witness to his greatness and renders us inexcusable before him. Studying creation is to understand the manifestation of divine power from the beginning of time, leading us to deep admiration and reverence.

Do you know what less than nothing is? I don't either, only God knows. He says in Isa 40:17: 'All the nations are as nothing before Him,' and look what follows. He regards them as less than nothing, as meaningless.

If you've never heard of less than nothing, then start believing now that nothing exists. This applies to everything. Even money. You have nothing in your pocket right now, but by faith, by the Spirit of God, you do. What you don't have, God brings into existence, coming from what doesn't exist, because he is mighty. To do much more, much more than we ask, think, or imagine, because his power is now working in us. For his invisible attributes, namely, his eternal power and divine nature.

The importance of studying creation is that creation reveals the glory of God. Man has become inexcusable before God because creation manifests God's glory and greatness. The creation of the cosmos is a preaching, a wordless, silent evangelization. Every time a man looks at the constellations, the stars, the galaxies, or studies them, he sees the manifestation of God's greatness and glory. Your power is being revealed in creation. So that every man becomes inexcusable. Man cannot come before God and say, Lord, I didn't believe because the gospel wasn't preached to me. The Lord will say, but the gospel was in the heavens, my beautiful, lovely, wonderful creation always revealed my glory and greatness!

For with God, nothing is impossible. This expresses the idea of nothing in an interesting way. For us, nothing is just emptiness. But for God, there are no impossibilities. He can do all things because he is God. Look at this. If, out of nothing, God created everything that exists, quantum mechanics raises some intriguing questions. Even when all particles are removed, there are fluctuations in the quantum vacuum of the universe.

Everything that exists is composed of particles: you, me, everything. Atoms, electrons, neutrinos, bosons, quarks—all these particles form the fabric of the universe. But even if all particles are removed, if everything is cleaned, leaving absolute emptiness, still, there are fluctuations. And what are these fluctuations? Previously, scientists struggled to understand how the universe could arise from nothing, as the Scriptures state. They thought the universe needed particles to exist, but now science has progressed.

The waves of fluctuation are a mystery. They come from a potentiality, an energy that has no particles, no tangible existence. This is a bold statement, but grounded in modern physics. It is a conclusion that shook the world of physics, but that the Scriptures had already indicated long ago. It took time, investment, and years of study to reach this conclusion, but it is a confirmation of what the Scriptures already taught. Perhaps it is not correct to affirm that God is antimatter, but I can affirm that God reveals himself through matter.

Universe from nothing means potentiality. This potentiality, we say we don't believe in the principle of the first cause, but we can affirm that God is beyond nothing. This is a very serious problem. All the mass and energy of the universe, all gravitational curvature, Einstein's theory are in this equation, all sums go to zero. Is this quantum mechanics? Is this quantum physics? The principle of singularity is zero, it's nothing, and the Scriptures say that God made the universe out of nothing. There's no escaping it, when you get to nothing there's no equation, no theory, it's not science, it's faith. Zero for God will be your starting point to add up what the result of an equation composed only of several zeros is.

Therefore, the Scriptures say that by faith you understand that what is visible comes from what is invisible by the word of God's mouth. If you take space-time, and compress it into a black hole, it will reduce to nothing. So, the idea of the word in singularity exists. So, the universe really came from nothing, originated from antimatter, arose from the vacuum as the Scriptures say. In the beginning, the cosmos, now visible, had no size at all. The big bang theory fails because there was never a point, there was never the starting point of an explosion, in the first second of calculation upon calculation, the universe never had any size. Zero for God is, the universe expands, but it always had the same size, it remains a zero point in the hands of God, no matter how big it is for man, the sum total of the universe will always be zero. But in the hands of the creator, it is an embryo without any dimension, large for man and small for God, because he is infinite and the universe finite, therefore, singularity is, because for God nothing exists, he said that the nations of the Earth are less than nothing to him. So, in God's hands, singularity is. For quantum mechanics it is not, it cannot go beyond singularity and created particles, it cannot comprehend, it cannot understand that there is something immaterial beyond a black hole, but for God, nothing and zero are ingredients for creation. And he said, "Let there be light," and there was light, nothing listens to the word of his mouth, because our

God is powerful! In the beginning, in singularity, we can put this initial principle of Genesis chapter 1, verse 1, at the beginning of creating the God of heaven and earth, as singularity. Nothing! There was nothing! There were no particles, no atoms, no quarks, no bosons, there was nothing, but in the beginning, in singularity, at the beginning of everything, God gave the start because everything was in God, but everything was not God.

He created the vacuum from where he began his creation. He created nothing, he created the vacuum, he created emptiness. The cosmos really "appeared", possibly, arose from nothing. I could hardly discredit the big bang theory. Any form will fall into the same equation.

Everything came from nothing, but nothing arose from nothing. In a time that was not time, the cosmos appeared, arose, expanded. Countless galaxies. Just in the Virgo cluster, which houses our Milky Way galaxy, there are more than a thousand galaxies, thousands of stars, and there are more planets than stars. Just a Virgo cluster. There are more than one hundred billion galaxies. Is that too much for you? It all started.

You came from nothing, from that tiny point, from that vacuum; there was your life, your marriage, your car, the guitar with which we play music—everything was in that small point. Everything boiled down to the primordial point, upon which God released his word; that point was nothing, or just a singularity, that quantum physics nothingness. We all came from that singularity or primordial point. The entire observable universe of two hundred billion galaxies, and if there are multiverses, the dead universe, which I believe exists, everything was in God, in the beginning, and he at the beginning of Genesis, when God began to create. I understand how Genesis begins by referring to the God of heaven and earth. This may encompass the Earth and correspond to all planets in all existing galaxies; planet Earth is just one among other Earth-like planets.

Many billions of earth-like planets scattered throughout the galaxies. Maybe not like ours habitable where God placed us, but Mars is an earth-like planet, Jupiter is an earth-like planet. So, by his power, God created many planets, including orphan planets, which are not around a star, if they are not visible with great ease, they are out there, if there are multiverses. The important thing is how everything arose and the Scriptures say that it was from nothing, by the word of God's mouth. So, this quantum fluctuation, which they talk so much about there, waves of fluctuation, in the vacuum universe, if you remove all the particles, it boils down to nothing. The universe from nothing means that it came from a potentiality, okay? All the mass and energy of the universe, all the gravitational curvature, the question of time and space, all the sums will give zero, okay? So, there's no escaping it in singularity. So the universe really came from nothing as the Scriptures say. This is a fact. This is quantum mechanics, it arrived at this conclusion, folks.

Everything came from nothing, but nothing arose from nothing. In a time that was not time, the cosmos appeared, arose, expanded. Countless galaxies. Just in the Virgo cluster, which houses our Milky Way galaxy, there are more than a thousand galaxies, with thousands of stars. And it is estimated that there are more than two hundred billion galaxies in the universe. Is that too much for you? It all started.

If someone comes with a contrary statement after discovering all the truth about the cosmos, it will be easily refuted. It would be a pity to waste time, effort, and money. We might even put an end to quantum mechanics and quantum physics. If someone comes with a contrary statement to the creation of the universe from nothing in the future. We can declare the death of quantum mechanics and quantum physics if it ignores the results of its investigations that have reached a conclusion. Billions of dollars were invested in building the James Webb to study the origins of the universe, but there is no surprise. After all, so many brilliant minds, present on our planet by divine will, dedicated so much time to reach the conclusion that the cosmos really arose from nothing, as the Scriptures suggest, after the final discovery to deny it suddenly, I can paraphrase Nietzsche. Science itself denies and equals: They are now suggesting that the universe be seen as an information processor.

Quantum physics is on the verge of an existential dilemma, where its own inquiry into the origin of the universe could lead it to intellectual demise. The discovery that the universe may have arisen from nothing challenges the foundations of physics and astrophysics. However, even in the face of this disturbing revelation, hesitating to accept this ultimate truth threatens the very investigative essence of quantum physics. By resisting this inevitable conclusion, there is a risk of obscuring the quest for truth and losing the ability to explore the deepest mysteries of the cosmos. It is crucial, therefore, for quantum physics to confront this uncomfortable truth and allow it to lead to new horizons of understanding, even if it means a radical redefinition of its established theories and paradigms.

Stating that the universe has always existed is akin to saying that the milk came before the cow. This issue will be explored in detail later on.

According to Scripture, there is no escaping this fundamental principle. In the beginning, the cosmos had no size whatsoever. Reflect on that. The Bible clearly states that there was no dimension, only absolute emptiness. It was nothing. Simply nothing. That is recorded in the Bible, isn't it? There was no form, no aspect. The Bible clearly shows us this truth, both in the Old and New Testaments. But in the hands of the Creator, singularity is significant. At the moment of singularity, as we have discussed before, everything was contained in God, under his dominion.

Reflecting on Jesus' role in the creation of the universe is crucial for understanding his magnitude and power. Just imagine: Jesus, facing a universe so vast, with its billions of galaxies and multiverses, all of that was insignificant to him. With a simple word, he brought into existence everything we know. The power of Jesus transcends any dimension, regardless of how we conceive of Him. His divinity is often questioned in the face of such grandeur. However, the truth is that all things, visible and invisible, were created through Him. It is intriguing how, after so much effort and dedication from brilliant minds to understand the origin of the cosmos, as described in Scripture, some now deny this truth. This sudden denial is already happening, and we need to address this issue. Some suggest that the universe be seen as a mere information processor. What mind is behind this exchange of information?

> "All things have been created through Him and for Him. The king of ages, immortal and invisible, the only God, deserves honor and glory forever. Amen" (1 Tim 1:17).

> "No creature is hidden from His sight; all things are exposed and laid bare before the eyes of Him to whom we must give account" (Heb 4:13).

> "I know that everything God does will endure forever; nothing can be added to it and nothing taken from it. God does it so that people will fear Him" (Eccl 3:14).

This is the ancient hymn of the Old Testament that Nietzsche tried to abolish, tried to extinguish, but it continues to be more alive than ever. Now, let's delve into George Lemaître's theory.

Understand: everything began with a primordial point. Although this point is indescribable, whether it exploded or not, everything expanded into a chaos of fire and multiple waves of energy. That's what the theory suggests, until it all became darkness, as the Scriptures mention.

And then, God said, "Let there be light." It aligns with Scripture. Let me explain. His theory starts from nothing, from this primordial point. Everything expanded, and there was fire.

Do you know what our God is? He is a consuming fire; his word is fire and life. There was nothing, and everything expanded, and the Scriptures describe in Gen 1:2 that there was chaos, darkness, and an abyss. Pay attention to what I'm going to explain; that's why it's important to read the Scriptures carefully. Genesis 1:2 tells us that God created the heavens and the earth; the earth was formless and empty, and the Spirit of God was hovering over the waters. Only in verse 3 does it say, "Let there be light," and there was light. When we study the Scripture text, we see that it is not concerned with the creation of the Earth itself in the following context. From the beginning, in Gen 1:1, it is mentioned that God created the heavens and the earth. Further on, the text highlights chaos and order, and then God says, "Let there be light!" An explosion causes chaos and disorder. There was an expansion of that initial pulse.

There is an abundance of scientific discoveries that support the idea of the big bang as the beginning of the universe. One example of this is the many evidences of helium that exist in the cosmos, along with radio waves and light that clearly indicate a beginning in this form. Therefore, when some suggest that the universe has always existed, following the information processing theory, it goes against decades of scientific research and efforts to validate such a fundamental theory.

The initial chaos expanded into a multitude of forms: into fire, energy, electromagnetic radiation, among others. There was a phase of rapid inflation, followed by gradual cooling. For about two hundred million years, the universe remained in complete darkness, according to some theorists. This description seems to align with biblical accounts that speak of darkness at the beginning of creation.

When God said "let there be light," as mentioned in the Bible, it can be interpreted as the beginning of the illumination of stars and celestial bodies, with the laws of gravity acting to organize everything as it should be. Some theories even suggest the occurrence of a supernova, a giant explosion that gave rise to the cosmos, which could explain the presence of trillions of comets around our solar system.

Although there may be disagreements between scientific explanations and biblical accounts, it is important to note that not everything can be completely reconciled. A careful study of the Old and New Testaments may offer some answers, but many questions remain open. However, it is undeniable that both science and religion offer valuable perspectives on the origin of the universe, and both deserve to be explored with an open mind and curiosity.

At the moment when God proclaimed the words "Let there be light," he was not laying the foundations of creation, but rather revealing the order that already existed. It was as if the universe, with all its complexity, was only waiting for that command to fully manifest itself. It is fascinating to realize how, even in the oldest Scriptures, we find reflections of the most modern scientific knowledge.

As the biblical narrative progresses, we see the process of creation unfold, from the initial light to the formation of animals, fish, and cattle. And later on, Genesis narrates again the creation of the heavens and the earth, emphasizing that everything was already present when God spoke the word. It was as if God had created everything first, and only then called for the manifestation of light; perhaps this explains the darkness and chaos. Everything was created before the Light; there was a hidden design that manifested all at once, but before that, there was an organized plan in creation.

The same Earth, which is 4.5 billion years old in God's time of creation, may have also manifested with thousands of years. This idea suggests the existence of two times in creation, something that perplexes me. However, I see no contradiction between science and a biblical interpretation free from religious dogma. I would be extremely happy if science is correct, but even more fulfilled if these two possibilities complement each other. I cannot conceive of a world without science, just as I cannot imagine a universe without God.

I want to go further. I believe that this known universe was brought into existence fully created to where it stands. God created the darkness and transported this universe, with all its creative activity of stars and galaxies, decreed by his word, to where it stands. Thus, the space that the transported universe occupied, which was left empty, forms the black holes that we are discovering. This explains the concentration of extreme gravity, since the laws of physics would not depend on matter to exist, as they were created before matter, as the Scriptures express in Genesis.

From this understanding, we solve the problem of black holes. They are just openings to a reality where our observable universe previously existed. However much I think, I cannot imagine a more integral solution to the black hole equation. The idea that this created universe was transported to where it is and left an equational void of darkness where it was, but still exists with the known laws, attracts stars and everything that dies in this universe to this dead dark dimension.

Therefore, God gave a command through fiat, which we will explain later, and all created things came from a place today called black holes, which must arise with the conflicts of gravity between an empty universe with nothing, but with the laws of gravity in full existence, and a new universe that God recreated everything for some reason, including bringing completely the so-called Pillars of Creation. They are three towers of gas and dust located at the heart of the Eagle Nebula. The first image of this fascinating cluster was taken by the Hubble telescope in 1995. The resulting image is among the most iconic space photos of all time. Surely, all planets, according to Gen 1:2, express that there was chaos on the face of the abyss, and the Spirit of God was hovering, making it clear a creative activity long before the structured creation in the poem of Genesis.

The theory of general relativity postulates that gravity is not just a force between massive objects, but rather a manifestation of the curvature of spacetime. When a massive star reaches the end of its cosmic journey and collapses under its own gravity, it triggers an event of cosmic proportions, opening doors to mysteries yet to be unraveled.

In this cataclysmic process, a region of such intense curvature is generated around the dying star that even light, the fastest messenger in the universe, is swallowed by its powerful influence. This dark zone, known as the event horizon, represents a threshold beyond which the known laws of physics lose their grip, plunging us into a realm of cosmic mystery and fascination.

If we conceptually think of an egg, it is made up of four main parts: shell, shell membrane, yolk, and white or albumen. The albumen would be the dimensions of the universe of yore, where God exercised all his creative activity, and where this whole solar system dwelled, including dinosaurs living and ceasing to exist in this other time and space, until for some reason it was transported entirely to where it stands but created about 13.8 billion years ago. It seems permissible to believe that the gravity of this other universe of highly incalculable dimensions has been pushing the galaxies away, a phenomenon we call the expansion of the universe known to us. This fabric of dark yolk exerts an undeniable influence on the albumen, which is formed by the same elements, but with the light color different from the yolk. This dead universe exerted an antimatter activity on our small universe.

An intriguing biblical text suggests that God dwells in darkness, perhaps this theory is an answer to the questions about the dwelling place of the almighty.

The theory of the dead universe better explains the theory of the expansion of the universe itself. We cannot in any way believe in an expansion from the explosion of the big bang; it makes no sense, for indeed the universe would decelerate at some point in its billions of years of existence. When I proposed the theory of a predecessor universe, where God somehow used the particles and all existing mass to create this observable universe, we find a more solid foundation for the theory of the expansion of the universe, which is undeniable. In fact, galaxies are moving away from each other, but not due to the expansion of a previous explosion, since there is not enough energy in the cosmos to cause continuous expansion. The gravity of the previous universe, as I explained in the yolk question, is attracting the observable universe back into itself; this strong attraction explains many phenomena that were previously complex for quantum physics to explain. This theory, soon, will certainly be elucidated in the light of scientific evidence, as research focuses on the study of black holes. When astrophysics discovers all the potentialities behind this divine architecture, we will certainly have a precise answer to the theory I am proposing in this treatise. Perhaps, on the other hand, when a star implodes and forms a black hole, this is also explained from the perspective of the gravity of this predecessor universe that exerts a strong influence on the formation of this strange phenomenon, which we can imagine as open crevices to the dead womb. Light, darkness, and mass have always been the proposal of the Scriptures for us to delve into the study of the cosmos from the perspective of a creative deity. In Gen 1:2, when it says: "And the earth was without form, and void; and darkness was upon the face of the deep. And the Spirit of God moved upon the face of the waters," we can make the analogy with the chicken egg that discusses the thing within the thing. This is explained by the composition, whether it is dark matter or any other type of unknown particle, it does not eliminate the issue of the hollow within the hollow, the previous void holding at its center the entire observable universe. The idea of the Genesis quote is the same when it states that a hen sits twenty-one days on the eggs until it hatches its chicks. The region where the Spirit of God incubated should be the previous universe, existing billions of years before this one known to us, extending its wings over all forms of the observable universe, creating planets, galaxies, comets, and everything else we can imagine. Therefore, astrophysics can affirm that the universe is expanding; on the other hand, affirming that this cause is still the energy generated by the big bang or by dark energy, which is proven not to exist in sufficient quantity in the universe to cause this expansion, would be like believing that the echo of the explosion of Hiroshima and Nagasaki could cause planetary chain destruction, as Robert Oppenheimer feared with the detonation of the atomic bomb.

The theory of the dead universe takes into account dark matter, radio waves, and particle behavior. My conviction that our universe is descended from its predecessor is supported by the biblical account of Gen 1:1-2. However, a creative agent does not negate the theory of the dead universe for purely scientific purposes. Rather, it should be the abode of the creator:

God dwells in darkness. I Kings 8:12: "Then Solomon said, 'The Lord said that he would dwell in thick darkness. I have indeed built you an exalted house, a place for you to dwell in forever.'"

Science does not strive to substantiate the existence of the divine; it merely seeks to investigate natural phenomena and elucidate them through the lens of empiricism. Similarly, it does not exist to deny the divine. Therefore, let us set aside that which escapes explanation and channel our energies into what can be explained—into the theory of the dead universe.

However, it is important to note that this is a highly speculative area of physics and cosmology, and currently, we do not have experimental or observational evidence to support the idea I am presenting. Developing new theories and conducting more detailed observations would be necessary to investigate this possibility more deeply.

The exploration of creative ideas is essential for the advancement of science, and this work moves in that direction of reflecting on the dilemmas and doubts of humanity. Even if my theory is not proven, the process of investigation and questioning must exist in the pursuit of understanding this complex universe.

Following my line of reasoning, I firmly believe that dinosaurs lived in the time and space of this universe before it became void and was transported into the reality we know for recreation according to Gen 1:1-2. In a logical sequence, the planets arrived first, for according to Genesis, they existed before light, and subsequently the stars were transported into this reality we know. Therefore, everything was in a state of chaos and destruction, as presented in the narrative of Genesis. There was chaos over creation and darkness before the existence of light. Thus, God decided to organize everything and prepare a universe for the existence of his noblest creation, human beings.

In eternity, there is no time, days, or hours. These concepts are human creations, limited by our finite understanding. Trying to express ideas about God the Creator using terms like "days of creation" is limited by human language. However, we will explore this issue further.

It is common to find criticisms of brothers in faith, the church, and Christians, often without foundation. This is evident even in academia, where a geologist sister was bullied for being a Christian. However, it is interesting to note that great thinkers and scientists who contributed to our understanding of the universe were indeed people of faith. From Giordano Bruno to Copernicus, many of the pioneers in scientific exploration were Christians.

Unfortunately, some people still maintain a narrow view of the relationship between faith and science. It is absurd to suggest that being a Christian and studying science are incompatible. The theory of the expansion of the cosmos, for example, was formulated by a man of God, Jorge Lamentri, whose work is widely recognized and underpins much of quantum mechanics and physics.

The cosmos is nothing to God, for to him, everything remains as a singularity, as quantum mechanics itself suggests, that everything is like nothing. Lawrence Krauss speaks about this, and it is true.

The theory of the creation of the universe is fascinating. According to it, from nothing, without energy, matter, or time, a tremendous explosion occurs that gives rise to space-time. From this explosion, a scorching chaos arises, followed by uncontrolled expansion, extending in all directions until it slowly cools. According to the Scriptures, God provided basic elements to bring into existence the entire expanse of the universe.

In the Genesis account, the "spirit of God" hovering over the void is mentioned, or as described in the Hebrew text, the "breath of God's lungs." This can be interpreted as the beginning of creation, where the echo of God's word resounded in the void, giving birth to the universe. God's speech produced powerful vibrations and waves that gave rise to creation. From the initial singularity, the cosmos began to expand in an instant, as mentioned in the verse.

This interpretation suggests that the creation of the universe was an extraordinary event, triggered by the word and voice of God, filling the void with life, fire, and energy. The expansion of the cosmos from this primordial point is a testimony to God's creative power.

It is sad to see how some people still perpetuate the idea that faith and science are in conflict. Albert Einstein, one of the greatest scientists in history, was a Jew who, although he had his doubts, showed his spiritual inclination throughout his life. The truth is that science and faith can coexist harmoniously.

In the beginning of everything, there is the Word that materializes, like a primordial force that gives rise to all creation. This is a fundamental concept that permeates both Scriptures and scientific understanding of the universe.

The idea that everything was created by the word of God resonates deeply, echoing the scientific understanding that the universe arose from nothing, from absolute emptiness. And just as the word of God brought light into existence, it also organized creation, giving shape and order to the primordial chaos.

The Word was there at the beginning, hallelujah! Glory to God, all things were made by him; without him, nothing that was made came into being (John 1:3). We cannot separate Jesus from creation, after all, in the beginning was the Word, and the Word was with God, and the Word was God (John 1:1). The angel of the church of Laodicea testifies that he is the beginning of the creation of God. By him all things were created, by him all things exist, for him and for his glory. As Col 1:15 says: "He is the image of the invisible God, the firstborn over all creation."

The word of creation is not a mere creation. God held nothing in his hands and proclaimed the word from his mouth, releasing powerful waves of creation. There was a progressive process in creation, as the Genesis text says. Although everything was processed at once, there were intervals of time. Therefore, the Old Testament text mentions days of creation. Although these days may represent significant intervals of time, for God, a day, a thousand years, or a million years are all the same.

Even when we study theories like that of the black hole, where a day in the black hole can represent millions of years outside of it, we realize the relativity of time. This is explored interestingly in films like "Interstellar." There was, therefore, a process of creation, as described in Genesis.

The word, it materialized in the form of atoms, of particles. It was like an active fire at the beginning of creation, and all that emerged in a second was the word from the mouth of God. If anyone doubted that everything arose suddenly, they must understand that I based it on the theory of a universe translated to where we observe it today. We Christians are very criticized when it is said that the Earth is a little over six thousand years old, but I believe in this theory, but if the universe was brought ready to where it is found? Why do I have to believe that the Cosmos arose in seconds and cannot believe that the Earth "appeared here" a little over six thousand years ago? Of course, I don't believe this because of science, but by faith, I don't deny this possibility either.

There was supervision in creation, as the Scriptures tell us. The Spirit of God hovered over the work of creation, generating life, like a hen hatching her eggs for twenty-one days. This is the same idea present in Hebrew: the Spirit of God is life itself, the divine breath. This supervisory work included meticulous quality control. When God created something, he proclaimed that it was good. When he created man, he declared it was very good, superior to all other creations. This shows that there was special care and quality control by God at each stage of creation.

Contrary to what some people think, we are not mere particles lost in an abandoned universe. God supervised us in creation and knows us deeply. He knows our thoughts, feelings, and desires. The Scriptures tell us that he even counts the stars and calls them by our name. We are creatures that brought satisfaction to the heart of the Creator.

It is important to understand that science is not in conflict with this idea of divine supervision in creation. On the contrary, when God gave us dominion over the Earth and the cosmos, he encouraged us to explore and study his wonderful universe. Scientific advances are a manifestation of this divine desire for us to dominate and understand creation. By using science to create technologies and explore the cosmos, we are fulfilling God's purpose for us.

Therefore, instead of criticizing science, we should recognize its importance and use scientific knowledge for the benefit of humanity.

From the beginning of creation, God created male and female, once again demonstrating his careful attention to creation. Although creation is subject to vanity, this is not due to God's will, but to the influence of other agents. It is interesting to note that both in Scripture and in the big bang theory, there is a moment when the cosmos comes into existence. This indicates the existence of a creative agent that preceded the initial singularity. The Latin expression *ex nihilo* reminds us that nothing comes from nothing. Only God, by his word, has the power to create something out of nothing.

When the Scriptures mention chaos upon the face of the abyss, we can consider that this abyss may be equivalent to black holes, upon which God worked and left traces for science to study and base its theories. For example, the detection of gravitational waves from the collision of black holes reveals not only a scientific achievement but also the wonder of divine creation.

We should marvel at the fact that God speaks to us through creation. While it is fascinating that science can detect cosmic phenomena, our true joy and satisfaction come from our relationship with God and our understanding of his glory and greatness as Creator. Whether there was an initial explosion or not, this is a question that God can clarify without the need for human interpretation.

Why was the universe created? He created without the need for interpretation. There is only a need to explain if there was an explosion, an expansion, like the big bladder, and we are within this cosmos. And there is a need to explain, so the theory of physics comes in, the scientists, that Christians follow the tradition of thinking scientifically, that physicists emerge, people who study quantum mechanics, to explain how everything was created. God created, period, the heavens and the Earth.

The laws of physics do not explain beyond singularity; this is the zero point from which God begins. So, to deny singularity for a universe that has always existed, that would be a denial of all the efforts of physics. Now, if it existed before as I explained, it may make sense, but the laws of physics cannot go beyond singularity.

We are creatures of God, in his image and likeness.

The idea that the universe has always existed denies all effort and investment, including the work of prominent figures like Einstein and the priest who created the big bang theory. Getting to nothing, to emptiness, is a frustrating perspective, but perhaps the universe has always existed, coming from another universe that no longer exists. Carl Sagan expressed this idea eloquently by stating that the cosmos is all that ever existed, that exists, and will always exist.

This perception leads us to reflect on the starting point of the big bang, where we encounter the wall of creation and singularity. If everything emerges from this point, from nothingness, it becomes complex and limiting. Quantum physics and other theories face difficulties in this context. The scientific community may resist the idea that the universe has always existed, but it is a possibility that cannot be dismissed. Perhaps God has only given a push to a universe that he recreated and will recreate and take to as many different occupations of time and space as he wants, since he is the creator and owner of everything.

The quest for answers leads us to the theories of new physicists, who challenge the idea of an absolute beginning, suggesting that the universe has always been present.

However, by embracing this perspective, we encounter a contradiction: all scientific progress, which has brought us closer to the truth, seems to be nullified. The Scriptures teach that the cosmos emerged from absolute nothingness, possibly referring to what is termed as less than nothing—a concept that challenges our most basic understanding. The citation of Isa 40:17's expression "less than nothing" suggests the idea that nations are considered as having no value whatsoever, or even a negative valuation, in relation to the infinite greatness of God. Although there is no explicit mention of subtraction in the mathematical sense in the biblical text, this expression implies a comparison where nations are viewed as a diminishment of the very concept of nothing, thus highlighting the disparity between their insignificance and divine supremacy. This poetic approach underscores the notion that nations are so minuscule and insignificant before God that their valuation can be interpreted as below zero itself, reflecting a true subtraction of value compared to divine magnitude. If scientists cannot explain nothingness itself, that is, what exists in quantum nothingness, then it is likely that they will have even more difficulty when the scriptures suggest the existence of a subtraction from nothingness itself, that is, less than nothing. Therefore, it becomes necessary to consider that, to comprehend these concepts, we may need to transcend our own human condition.

Understanding the universe as an information processor offers an intriguing view, especially in light of religious narratives. However, rejecting all accumulated knowledge in favor of this approach would be imprudent. After all, if the universe has always existed, why would it need to expand? This contradiction brings us back to the idea of a primordial singularity, the zero point from which everything originated, as the Scriptures suggest.

The understanding that God created the universe through his word is fundamental to understanding not only the physical nature of the cosmos but also its relationship with spirituality and human life. It is important to recognize that the Scriptures offer not only spiritual guidance but also a basis for scientific understanding of the world around us.

The understanding that God is the Word and that all creation was brought into existence through it leads us to a profound reflection on the relationship between divinity and the universe. By recognizing that God's word is the spirit and life that was present at the beginning of creation, we are invited to delve deeper into the search for truth about the origin and nature of the cosmos.

The connection between God's word and creation is evident not only in Scriptures but also in various scientific theories. The biblical account offers a unique light on events like the big bang and other scientific discoveries, suggesting that many of the fundamental concepts of physics and cosmology have their roots in the divine word.

As we confront the limitations of human knowledge in the face of singularity and concepts like nothingness and less than nothing, we are challenged to transcend our limited understanding and recognize God's presence in all things. In this sense, God's word not only guides us on our spiritual journey but also offers profound insights into the nature of reality.

Ultimately, by recognizing that everything was created by the word of God and that he is the foundation of all existence, we are invited to render him honor, praise, and reverence. Understanding that we have everything in God allows us to face the challenges of life with confidence and gratitude, knowing that he is the consuming fire that guides and sustains us in all things.

It is essential to remember that, regardless of scientific theories or religious interpretations, the mystery of creation remains. The Genesis account transports us to the moment when the Earth was without form and void, when the spirit of God hovered over the waters. This poetic image reminds us that, from the beginning, God imprinted his word on matter, sustaining all visible and invisible things. And so, even in the face of uncertainties, we trust that God does not act without his word, guiding the steps of the universe from its origin to the present day.

# Chapter III

## Vision of a Universe without God

In exploring the vast cosmos, some contemplate a universe devoid of God, a realm governed solely by the laws of physics and chance. In this vision, the universe emerges from primordial chaos, guided only by natural processes and devoid of divine intervention.

In this perspective, the big bang serves as the starting point, a cosmic explosion that birthed space, time, and matter. From this initial singularity, the universe undergoes expansion and evolution, driven by fundamental forces and governed by mathematical principles. Over billions of years, stars form, galaxies collide, and planets emerge, all within the framework of natural laws.

Life arises through a process of chemical evolution, where simple molecules give rise to complex organisms through a series of incremental steps. Through the mechanism of natural selection, life diversifies and adapts to its environment, leading to the emergence of diverse species.

Human consciousness, once seen as a divine spark, is viewed in this vision as a product of biological processes, arising from the complexity of the brain. Our thoughts, emotions, and experiences are the result of neuronal activity, shaped by evolutionary pressures over millions of years.

In this universe without God, morality is a human construct, shaped by societal norms and cultural values. There is no absolute right or wrong, only subjective interpretations of ethical behavior. The quest for meaning and purpose is a personal journey, devoid of divine guidance or ultimate destiny.

Yet, amidst the vastness of space and the complexity of life, some find a sense of awe and wonder. They see beauty in the intricacies of the cosmos and find solace in the interconnectedness of all things. While they may not believe in a personal deity, they acknowledge the mystery and majesty of the universe.

Whether one sees the universe as a manifestation of divine creation or as a product of natural processes, the quest for understanding and exploration remains. Whether guided by faith or reason, humanity continues to seek answers to the fundamental questions of existence, driven by curiosity and a desire to unravel the mysteries of the cosmos.

In this chapter, we will continue to explore the theme of creation, confident that God will strengthen our faith through the arguments presented in this work. It is part of a series of teachings that address the roots of biblical Christianity, grounded in the Torah, likely written by Moses, according to the sages of Judaism and Christian theologians. We face opposition both internally and externally. Internally, there are challenges regarding the interpretation of the first three chapters of Genesis, which form the basis of our faith. If these chapters were excluded from the Scriptures, we would be lost as a church, for we believe that God is the creator of the heavens, the earth, and all that is in them.

Previously, we explored advanced concepts of quantum physics and quantum mechanics, emphasizing the principle of creation ex nihilo, where God brought something into existence out of nothing. Although our intention is not to reconcile science and Scripture, but rather to present this possibility in an argumentative manner, it is important to submit scientific concepts to the authority of the Holy Scriptures, maintaining the integrity of biblical theology. On the other hand, we apply scientific concepts to demonstrate the relevance of science in the eyes of Scripture scholars. We also face external challenges, such as liberal theology, which flirts with evolutionism merely out of vanity and questions the traditional interpretation of the Scriptures. Darwinism, for example, does not offer satisfactory explanations for the origin of life, unlike the Scriptures, which assert that life comes from God.

Science plays a crucial role in understanding natural phenomena, including the creation and study of the cosmos. However, some scientific currents advocate for the possibility of a universe without divine intervention, challenging the fundamental principles of biblical creationism.

Many scholars, who once professed the Christian faith, now adopt theories that directly question the biblical narrative of creation. When faced with the difficulty of convincingly refuting biblical creationism, they resort to creating new theories, such as the multiverse and eternal inflation, as if they existed to deny creationism.

The Scriptures affirmed millennia ago that the universe was created out of nothing, long before the development of these scientific theories. Even with the existence of multiverses, the question of the origin of the universe remains without a definitive answer, inevitably leading us to singularity.

It is time for the church to take a stand and understand science to face these challenges. The frustration of atheists upon realizing that a priest contributed to the theory of the big bang reflects the difficulty in finding a complete scientific explanation for the origin of the universe that denies divine creation.

The theory of the big bang, although proposed by a Christian scientist, was not created to refute the Scriptures, but to systematically understand the process of creation. All scientific theories, fundamentally, recognize divine creation. It is necessary to advance in the understanding of these issues, aligning them with orthodox and reformed biblical theology. It is time to synthesize these teachings to strengthen the faith and understanding of the readers.

So, let's read a few more verses that go like this: "In these last days he has spoken to us by his Son, whom he appointed heir of all things, and through whom also he made the universe" (Heb 1:2). I reaffirmed that the universe, all that exists, comes into being through Jesus or Yeshua, as some prefer, by the word from God's mouth and being Jesus himself the word. Also being the breath of God that is there in Genesis chapter 1:2, which says that the Spirit of God was hovering over the face of the waters, over the earth, the breath of God, the Spirit of God, the life of God, the lung of God. To please the theorists of the big bang, he blew the balloon. He blew life, the Spirit, and everything happened and came to be.

For every house is built by someone, and God is the supreme builder of everything (Heb 3:4). God is the creator in Genesis chapter one, verse 1:1 and Genesis chapter 1:3, doesn't exist to prove the existence of God, but to bring God as the evident creator of all things. A God who needs to prove his existence, indeed, is not God. He doesn't need to prove anything. And what did God do? What is the greatest proof of the existence of God? It's Jesus incarnate, God made man. He manifested himself, walked, lived as a man, preached the gospel, said he is the son of God, God himself, he and God are one, but men don't believe in that. Trying to prove the probability of the existence of God is something totally irrational and absurd.

The atheist must know that God incarnated himself and became a man. He humanized himself to save all who believe in him. He died and rose again. This God is composed of Father, Son, and Holy Spirit, according to the Scriptures. So, trying to discover a God through science, through physics, if they actually succeeded, it's not the true God, it's not the God in whom we believe. Our God, his thoughts are above our thoughts. Our God, he is the creator, we are creatures. The human mind cannot fathom the infinity of God, cannot fathom the heart of God. So, seeking the existence of God only through methods, scientific ideologies, is an absurdity, it's intellectual despair, it's intellectual death. The Scriptures point to a God in evidence who created the heavens and the earth, and this God became flesh and dwelt among us. He brought the good news of the kingdom, Jesus preached the gospel, was nailed to a cross, died and rose again among the dead, as the Scriptures affirm.

According to physics, there are around seven dimensions, including time and space. God is not bound to them, he created all these things. Genesis chapter 1:1 presents him as the creator of the heavens and the earth, as well as of all dimensions, including the time and space that we currently know through the study of physics.

> "So says God, the Lord, who created the heavens and spread them out. He extended the earth and all that sprouts from it, giving breath of life to the people who dwell in it and spirit to those who walk through it." (Isa 42:5)

In the context of string theory and the understanding of time and space, we can imagine the heavens being depicted as a book, a scroll that rolls up. The book of Revelation describes the heavens rolling up like a scroll, indicating a moment of cosmic transformation.

> "The sky receded like a scroll being rolled up, and every mountain and island was removed from its place." (Rev 6:14)

> "All the stars in the sky will be dissolved, and the heavens rolled up like a scroll; all the starry host will fall like withered leaves from the vine, like shriveled figs from the fig tree." (Isa 34:4)

For God, the cosmos is like the scroll of the Torah, which he unfolded when releasing his word and creating. He unfurled the heavens before us by the power of his word and the breath of his mouth, by his Spirit of life.

It is crucial to highlight that our church is grounded in a serious, Reformed, and biblical theology, where we do not consider the possibility of flirting with theological liberalism. The decline of churches in the United States is an alarming symptom of the gradual removal of biblical theology from their foundations. The crux of the matter lies in the denial of the infallibility of the Scriptures, which are not only the source of faith but also the supreme guide for religious practice. The centrality of God, his glory, honor, praise, and worship, as well as salvation exclusively through Christ Jesus, are non-negotiable principles that uphold genuine Christian faith.

When some voices begin to question the integrity of the Scriptures, suggesting that they contain human errors and cultural influences, firm foundations are shaken. This is the beginning of the decline of American churches. However, there is a glimmer of hope emerging from this theological darkness. A rescue movement is taking place, mainly in the Southern Baptist church, which historically has been one of the largest religious denominations in the United States.

Ministers and leaders are recognizing the urgent need to return to the fundamental principles of biblical theology and Reformed theology. This rescue is not just a restoration but a renewal, a reaffirmation of the values that have shaped Christianity over the centuries. This approach rescues Christian faith in a more intense and profound way than any other contemporary theology.

In the beginning, God created the universe not through primordial particles or pre-existing elements, but by his spoken word. God released his word, and by the simple divine declaration, all things came into existence. While theological liberalism and even some currents of Jewish thought attempt to explain the origin of the universe in material and scientific terms, the biblical truth is clear and unshakeable.

God did not need raw materials to create; he is the very source of creation. There were no particles, quarks, or atoms before the word of God echoed in the cosmic void. The universe did not emerge merely from a primordial soup without God's word but was brought into existence out of nothing by God's powerful word.

This fundamental understanding of divine creation not only reaffirms the sovereignty and power of God but also challenges materialistic and naturalistic conceptions of the origin of the universe. It is by the word from God's mouth that all things were created, and it is in this truth that we find the solid foundation for our faith and understanding of the cosmos.

The biblical account of creation is clear and undeniable. God, through his word, brought all things into existence. When he said, "Let there be light," light emerged. This narrative transcends the limitations of human understanding, challenging us to accept divine supremacy over creation.

In contrast to this fundamental truth, some seek to subordinate biblical theology to science. However, the essence of biblical faith rests in the conviction that God created the universe by the word of his mouth. He spoke, and creation obeyed, manifesting according to his will.

It is crucial to recognize that everything that exists was created and approved by God. Nothing is outside of his sovereign plan. Faced with the difficulties and chaos we often witness in this world, we must understand the true expression of God's love. He not only created everything according to his desire but also allows history to unfold according to his will.

Every detail of our existence, from the seat we sit in to the camera that observes us, is part of the divine plan. Accepting this reality is embracing God's sovereignty over all things, even when we do not fully understand the events that occur around us. Everything that has happened in the past and everything that is yet to come is within God's designs, and any deviation from this plan is contrary to his will.

The current theological approach represents a threat to spirituality, leading to spiritual death. God chose to reveal himself through his works, not through direct references to himself. When he manifested himself in Gen 1:1–3, he first revealed his works, his greatness, his majestic power, and his sovereignty over all creation. He revealed himself as the almighty, the omnipotent, the omniscient, knowing all things, demonstrating his wisdom in all the work of creation.

God created all things in a planned and organized manner over a period of six days, as the Scriptures state. However, it is important to recognize that the counting of days only became relevant to humanity after the fall, when mortality entered the scene. Before that, humanity lived in a plane of eternity, making it impossible to determine the duration of the time Adam lived without sin.

While it is challenging for some to understand the biblical narrative within the context of modern science, the Bible clearly states that God created everything in six twenty-four-hour days. This does not deny God's power and greatness, but, on the contrary, validates his authority over all creation. It is through this perspective that we can fully appreciate God's sovereign control over time, space, and all dimensions of existence.

As we reflect on creation, we are not concerned with the exact age of the Earth or the universe, as modern science suggests. Instead, we focus on the fundamental truth that God is the supreme Creator, and all creation is under his dominion and care. It is this understanding that allows us to recognize and glorify the greatness and power of God.

The discussion about the age of the Earth and the Cosmos is irrelevant to genuine Christianity. For us, what is important is to recognize and accept the biblical narrative that God created everything in six literal twenty-four-hour days. This understanding not only exalts the power and greatness of God but also preserves the integrity of Scripture as the ultimate authority in matters of faith and practice. Any deviation from this literal interpretation of Scripture represents a danger of entering into liberal theology.

While we acknowledge the use of anthropomorphisms in the language of Scripture to express ideas about divinity, it is vital to maintain an approach that respects the literalness of the creation account in Genesis. God created in six days, as the Scriptures affirm, but man was not present to witness these events. Science, in studying fossils and traces of Earth's history, may provide valuable premises, but it cannot determine the exact chronology of creation.

The discussion about dinosaurs is also complex and often misunderstood within the context of faith. Some theological lines suggest that dinosaurs coexisted with Adam and Eve before the Fall, but it is important to recognize that the rebellion of animals occurred after human transgression. In the garden, there was peace, except for the presence of the serpent, which represented evil and spiritual downfall.

By committing ourselves to a faithful interpretation of Scripture, we acknowledge God's sovereignty over all creation and avoid the danger of distorting his Word to fit human ideas. Understanding creation in six days leads us to a renewed reverence for the power and majesty of God, which is above all human understanding.

The issue of dinosaurs is often surrounded by speculation and various scientific theories, but it is important to address it within a consistent theological perspective. The most accepted theory suggests that the extinction of dinosaurs was caused by an asteroid impact, diverted only a few centimeters from its original trajectory. However, I lean towards believing in a more theological interpretation, which considers Satan's fall as the event that led to the extinction of dinosaurs and other prehistoric animals.

This theological position stems from the understanding that spiritual death, symbolized by Satan's rebellion, is the true beginning of the separation between man and God. I believe that Satan's fall caused devastation in God's creation, leading to the extinction of dinosaurs and other forms of life.

When the serpent tempted Eve in the Garden of Eden, it was not just a snake, but the manifestation of Satan, a spiritual being who needed a physical body to express himself. Similarly, Jesus' incarnation through Mary was a way for God to legally enter the physical world, showing that Satan also seeks this physical manifestation to fulfill his evil purposes.

The demonic possession of animals, as mentioned in the Scriptures, illustrates how Satan seeks to use physical bodies to express his wickedness. By understanding this spiritual dynamic, we can see the extinction of dinosaurs as part of a larger plan by Satan to corrupt God's creation and divert humanity from its original purpose.

This theological interpretation may not be accepted by everyone, but it is important to consider that the issue of dinosaurs does not alter the fundamentals of the Christian faith. As we address complex themes like this, we should maintain a posture of humility and openness to different perspectives, always keeping our focus on the centrality of Christ and the authority of Scripture as our spiritual compass.

The introduction of death in the garden was a crucial moment in human history. It was through Adam and Eve's disobedience that spiritual death entered the world, separating man from God. Physical death, as we know it, is only an outward manifestation of this spiritual separation.

The theory of Satan's fall as the cause of the extinction of dinosaurs may seem controversial to some, but it is important to understand that it fits within a broader theological perspective. The presence of fossils and evidence of extinction leads us to reflect on the role of death in God's creation.

The recreation of the Earth described in the Scriptures leads us to understand that God did not create it to remain empty but to be inhabited. The Earth was formless and empty until God brought order and life through his Word and his Spirit. This theological view aligns with the principles of the Torah and leads us to recognize that science is subject to divine authority.

Professor Nathan Aviezer, a Christian scholar, observes that science is in harmony with the teachings of the Torah. Ultimately, it is the Word of God that has authority over all things, and it is to it that we must submit in our quest for understanding and knowledge.

The universe was conceived out of nothing, as the Scriptures affirm. In this sense, any scientific discovery, whether in quantum mechanics or any other area of physics, must submit to the authority of the Scriptures. The fundamental truth is that the heavens and the earth were created out of nothing, and this needs to be recognized. We are not seeking harmony between science and Scripture, but rather the recognition that the latter represents the ultimate truth.

God created the universe out of nothing, and all scientific discoveries, made possible by the intelligence granted by God himself, must be valued. Every scientific advancement reveals the glory and greatness of the Creator, demonstrating the capacity and intelligence that he has shared with us. When we consider even events like the big bang, we see that the biblical account of Genesis 1, where God creates the heavens and the earth by his word, does not conflict with scientific discoveries.

When addressing the creation of the universe, it is important to understand that God did not create from something pre-existing, but from absolute nothingness. He spoke, and light appeared, bringing order to chaos. Even though contemporary physics suggests different scenarios for the origin of the universe, the biblical account remains the supreme authority. It is essential to synthesize these concepts to deeply understand the relationship between science and faith in explaining the origin of the universe.

The creation of light is seen as the essential moment of the creation of the universe, something that the sages of Judaism and Christianity affirm to be in total harmony with the latest scientific discoveries.

It is important to understand that recent scientific discoveries fit within what the Scriptures stated long ago. It is not that the Scriptures fit within scientific theories, but rather the opposite. The word of God, expressed in the Torah, is seen as the ultimate truth, and science, through research and study, eventually arrives at the same conclusions that the Torah proclaims.

This convergence between Scripture and science is not extraordinary, but rather a confirmation that Scripture was correct from the beginning. Even those who do not hold religious beliefs are confronted with the evidence that Scripture and science are aligned in affirming that God created the cosmos out of nothing.

Scientists point to an explosion known as the big bang, which has been discussed previously, as a possible cause for the origin of the universe. However, the crucial question remains: what caused this explosion? From a scientific standpoint, there is no definitive cause identified. Some scientists even fear that accepting the existence of God would mean the end of their intellectual careers.

It is interesting to note that the theorist Georges Lemaître, who was a Christian, was responsible for developing the theory of the big bang. Einstein, on the other hand, oscillated between his Jewish beliefs and his doubts about the existence of God, reflecting a duality that many face when approaching these questions.

Scientists point to the idea of an uncaused explosion or the spontaneous creation of the universe, a term widely discussed today. This notion may seem absurd, but it's as if suddenly there was a rain of hamburgers in your house without any apparent scientific explanation. The idea of spontaneous creation suggests that the universe arose from nothing, without any logical explanation within the context of known physical laws.

Reflecting on spontaneous creation can lead us to question the very foundations of physics and gravity. It's as if we're faced with a scene from a movie where hamburgers start falling from the sky for no apparent reason. This conception challenges even our capacity for comprehension and belief in the world around us.

Indeed, sometimes it seems easier to believe in something as absurd as the idea of spontaneous creation than in biblical narratives. Imagining a machine that makes it rain hamburgers seems entirely out of the question, while the notion that the universe arose from nothing, spontaneously, is something that some try to defend. It is simply irrational and defies any logic.

Jews and Christians defend the idea that it was God who created the Earth out of nothing, simply with his word. The moment he said "let there be light" was the moment of creation. For us, it is clear that there was a cause, and that cause was God. But for some scientists, it is easier to attribute the origin of the universe to spontaneous creation, without any causing agent.

However, it is interesting to note that all scientists agree that there was an initial moment when the universe arose. It is a matter of interpretation whether this origin occurred through divine action or through a spontaneous explosion. Some prefer to believe in the latter option, while others recognize the wisdom behind the biblical narrative.

So, who created the big bang? This is a question that can be approached both by science and theology. The Scriptures affirm that it was God who created the cosmos out of nothing, simply with his word. And that's what we're exploring here: the idea that God shaped and organized the universe according to his will, as described in the Scriptures.

Recently, science has made significant discoveries, such as the detection of cosmic background microwave radiation, which permeates the entire universe, and the understanding of the so-called dark fabric, a weak radiation present everywhere. These gravitational waves leave traces as they traverse space during the inflationary phase of the universe, functioning as an echo of the big bang, as suggested by the scientist and Christian Lemaître.

This scientific explanation, which aligns with the biblical account of Genesis, should be subjected to the authority of the Scriptures. Any deviation from this principle is considered theological liberalism. The Bible clearly states that God created everything out of nothing, by his word. As Isa 40:12 reminds us, who can measure the waters in the hollow of his hand or weigh the Earth on a scale? It is God who, with wisdom, founded the earth and prepared the heavens.

Scientists, like John N. Kovach of the Harvard Center for Astrophysics, have made notable progress in this understanding. Recently, the BICIP2 project team announced the discovery of residues left by cosmic inflation, which reinforces the idea that the universe had a beginning and was created according to biblical accounts.

The name given to the exponential growth through which the universe passed in its first quadrillionth of a second is known as cosmic inflation. This phenomenon opens a window to a new realm of physics, revealing truths about the events in the first fraction of a second of the universe. Kovach, the leader of the Bicep2 project team at Harvard University, emphasized the importance of this study.

Every scientific attempt to understand the cosmos always returns to the idea that the universe arose from nothing. By studying theories and evidence, scientists reach conclusions that corroborate what the Scriptures have been affirming for centuries. These discoveries not only open a new perspective but also encourage scholars to recognize that the Scriptures were right.

The chronological order of creation, as described in the Scriptures, is interpreted literally. The days are considered periods of twenty-four hours, and light is the first created element, as indicated in the biblical accounts. This understanding encourages deeper investigation from the perspective that Moses wrote a treatise that encompasses geology, astronomy, and physics in an accessible manner. In this context, all science must submit to the authority of the Scriptures, as they represent God's truth.

It's an interesting question. Why is science that postulates that the universe arose from nothing considered scientific, while Scriptures expressing the same idea are labeled as religious tales? Moses, through his divine inspiration, wrote a treatise that addresses not only religious matters but also historical and scientific subjects.

When it comes to creation, the biblical narrative must be seriously considered as a historical and scientific account. While science proposes theories and studies about the origin of the universe, often based on solid foundations, the idea that the universe arose from nothing and that life on Earth was the result of chance also requires a significant amount of faith to be accepted.

The approach of Stephen Hawking, for example, suggests that the laws of physics are responsible for the creation of the universe and life on Earth, implying a universe that arises spontaneously, without the need for a divine creator. However, this view raises questions about the meaning and origin of the laws of physics and the complexity of the universe itself.

Therefore, it is important to recognize that both science and the Scriptures offer perspectives on the origin of the universe and life, and each can be examined in light of its own knowledge base and belief. The biblical narrative is not just a tale but a progressive revelation of God's truth to humanity, addressing not only spiritual but also historical and scientific issues.

It is difficult to conceive that the laws of physics and gravity are responsible for the creation of the universe, especially when it implies denying the existence of a divine Creator. The renowned physicist Stephen Hawking, for example, once considered the possibility that God created the universe, but later developed a theory suggesting that the laws of physics are sufficient to explain its origin.

My suggestion is that the laws of physics were established by God before the formation of the visible universe; they merely repeated in this cosmos. He instituted these ancient laws in this new universe to shape it before bringing all matter into it. This conception highlights that we cannot comprehend the universe merely by the existence of matter, for there is an immaterial universe governed by laws that influence this known universe. We cannot examine the visible universe without considering the possibility of another preceding one; the biblical texts provide permissive evidence for believing in this other reality.

> "The earth was without form and void, and darkness was over the face of the deep. And the Spirit of God was hovering over the face of the waters" (Gen 1:2).

The existence of water at the beginning of this activity outlines the plan for the existence of life and has become one of the greatest challenges of science: finding water on other planets. The earth and everything we can observe are within a large black hole; we are living inside it with everything we know and still unknown. We cannot deny this according to the Scriptures. The probability of an older universe reinforces my idea that we are inside a black hole embedded within another black hole, and who knows if the universe is not an overlap of black holes wrapped in black holes, forming a large roll of these realities, each with its own physical laws, exerting influence on each other. Why not believe that what exists is not only what we can see with our eyes or study with a telescope? The laws of gravity exist and are laws of gravity, but every form of life we know always encounters the existence of matter. What is life? A combination of biology with energy and other components of the universe added to a thinking brain? Can all life only be life if it is matter? Why can't there be realities with immaterial life? What is the absurdity surrounding this issue? I do not conceive the existence of black holes without the existence of a universe from which our universe originated, but which today is completely governed only by the activities of the laws of physics, such as gravity, and perhaps other laws unknown to us. I find the explanation that star explosions and supernovas cause black holes to be very superficial, although I do not deny that it may be a probable truth.

While my theory is not yet supported by observable and testable scientific evidence, it cannot be dismissed like other speculative theories. If we can consider the possibility of multiple universes or parallel universes, where the laws of physics may vary between them, then my theory that we are inside a small black hole, which may even be central, like the roll of the book, involved in multiple black holes, which make up this total fabric, cannot be dismissed as mere speculation.

This idea is explored in theories such as string theory and the theory of multiple worlds, while I propose black holes with their own laws as if they were universes. On the other hand, I believe that there is only one universe with the nature like our universe. The other realities are just black holes composed with known and unknown laws that influence our cosmos. In these scenarios, it is conceivable that our universe is unique, surrounded by many black holes, since it resides within a black hole. We can say that each of the other "universes," without light, matter, earth, but with unknown energies, exists, with their own physical laws. I go even further to think that unknown elements within this universe may have come from these other realities, such as dark energy, which defies our conventional physical laws. However, these theories remain speculative and currently lack direct empirical evidence. In the context of my idea, it is interesting to note that it shares certain similarities with these speculative theories, as it also suggests the existence of a reality beyond our known universe, with its own physical laws.

However, it is important to distinguish between speculation and empirical evidence. While my idea offers an intriguing and philosophically rich perspective, it is still not supported by observable and testable scientific evidence. However, like other speculative theories, it cannot be dismissed without further investigation.

> "The earth was without form and void, and darkness was over the face of the deep. And the Spirit of God was hovering over the face of the waters" (Gen 1:2).

> "Thus says God, the Lord, who created the heavens and stretched them out, who spread out the earth and what comes from it, who gives breath to the people on it and spirit to those who walk in it" (Isa 42:5).

> "The sky vanished like a scroll that is being rolled up, and every mountain and island was removed from its place" (Rev 6:14).

When I focus on these verses, I conceive a state prior to the known creation and existence in the vacuum from which God formed all dimensions of black holes. What I can assert is that the small black holes seen by science through telescopes are only tiny perceptions of a larger reality. Whether they are formed by explosion or not, this does not invalidate this argument, as I am talking about something beyond scientific observation, yet entirely possible.

The richness of detail in these biblical texts of few words impresses me, as the author suggests that before the presence of physical elements like particles, earth, and water, there was a black hole called the Great Abyss. This cannot be anything other than the small black hole where the universe we live in resides. The existence of water at the beginning of this activity outlines the plan for the existence of life and has become one of the greatest challenges of science: finding water on other planets. The Earth and everything we can observe are within a large black hole; we are living inside it with everything we know and still unknown. We cannot deny the probability of this truth according to the sacred Scriptures. There must be an older universe where everything originated, different from ours; everything that exists and can be observed was created in another universe and was brought completed to this young universe, leaving the other empty where the predecessor universe was. God first created the laws of physics in this known universe, just as in the old universe the laws of physics must continue to exist, causing a strong influence on our universe. This reinforces my idea that we are inside a smaller black hole incorporated within another black hole of larger dimensions, called the primordial universe. This premise to become understandable used the metaphor of the egg of a chicken. Perhaps the universe is an overlap of black holes wrapped in black holes, forming a large roll of these dimensions and various realities that will soon be proven by quantum physics and cosmology. Each of these dimensions of black holes has its own laws that govern them, exerting influence on each other and causing distortions in the space-time of our young universe.

In quantum mechanics, there is the concept of quantum fluctuations in the vacuum, where particles virtually arise and disappear rapidly due to quantum uncertainty. These phenomena suggest that the quantum vacuum, or "nothing," is not truly empty, but rather a sea of quantum activity as I stated earlier. Can a visible universe emerge from an invisible universe where no particles exist?

In light of these concepts of quantum physics, my idea that the laws of physics existed before the creation of the visible universe and were dimmed to shape it can be challenged, but cannot be refuted. The universe may emerge from a fundamental quantum state by the vacuum of God's voice, where he, as the lord of the laws of physics, can intertwine all these previous dimensions with the existence of the universe itself. The idea of a universe within a black hole, which I believe can be the center of activity of all other existing black holes, which exert influence on this one, and which also receive influence, such as when there are explosions and small black holes are created, leaves me perplexed in light of biblical quotations like these that seem to make my assertion clear:

> "All the host of heaven shall rot away, and the skies roll up like
> a scroll. All their host shall fall, as leaves fall from the vine, like
> leaves falling from the fig tree" (Isa 34:4).

My sincere question to you is why must a book have blank pages and cannot exist as a roll of black hole sheets rolled one over the other, and what you need to know is that in the context in which these quotes were written, books were on scrolls of papyrus that they would unroll to study the Scriptures.

In my theory, the laws of physics existed before the creation of the visible universe, and these laws were dimmed in this universe to give it shape merit additional credit. Since a more absurd idea proposed by Stephen Hawking, that the laws of physics created the universe, seems far from the reality of quantum physics.

In quantum physics, there are concepts like uncertainty and the principle of superposition, which challenge our traditional understanding of causality and the determination of physical laws. For example, the principle of superposition suggests that a quantum system can exist in multiple states simultaneously, until it is observed or measured, at which point it collapses to a specific state. This implies that the properties of the universe may not have an objective reality before being observed.

Why not believe that what exists is only what we can see with our eyes or observe with a telescope? Gravity laws exist and are just gravity laws, but every form of life we know always encounters the existence of matter. What is life? A combination of biology with energy and other components of the universe summed up with a thinking brain? Can all life only be life if it is materialized? Why cannot there be realities with immaterial life? What is the absurdity around this issue? I do not conceive the existence of black holes without the existence of a universe from which our universe originated, but today is completely governed only by the activities of the laws of physics, like gravity, and perhaps other laws unknown to us. I find it very superficial the explanation that star explosions and supernovae cause black holes, although I do not deny that it may be a likely truth.

Can God not exist? I find it unlikely. On the other hand, God's existence in the universe implies that some questions will never be answered without knowing his mind deeply. On the other hand, many people, especially those with religious convictions, have difficulty accepting this view of a universe without God. In my particular case, nothing will be added or taken from my life with this absence of God in the universe because, in this case, without him, life is nothing, while on the contrary, we will always have something to lose. For them, the idea of a cosmos without a divine cause is desperate and disappointing, implying that life and the universe itself are the fruits of chance, without purpose or transcendent meaning.

It is understandable that these theories that point to a universe without God generate anguish and despair for those who believe in the existence of a Creator. After all, if the universe and life arose by chance, what would be the purpose of our existence? These are deep and complex questions that challenge our understanding and lead us to reflect on the meaning of life and human existence.

The laws of physics explain existence through gravity, suggesting that we came about by chance. For some, this implies a meaningless void, where our life and efforts have no meaning, as there is no divine Creator who gave purpose to things, as the Scriptures affirm.

This post-modern perspective of an absurd, empty, and meaningless universe is gaining ground, spreading total disbelief. Many who once professed faith now abandon it, embracing the idea that the universe and life arose from nothing, without the need for a creating God.

It is interesting to observe how this view contrasts with the belief in a creating God. Some claim that the universe arose from the laws of gravity, without the intervention of a divine being. It is an approach that challenges the traditional conception of divine creation, suggesting that the universe self-created through physical laws that, for some, are mere nothingness.

The expression *ex nihilo* reminds us that nothing comes from nothing. It is a statement that resonates with the idea that God is the creator, not nothing. Then arises the question: how to explain the emergence of the universe from the laws of gravity? Who created these laws?

When faced with this question, the scientist finds himself at an impasse. He suggests that the universe self-created from the laws of physics, but does not offer an explanation about the origin of these laws. This raises doubts about his theory, after all, one cannot assert that the universe arose from nothing without first considering who created the laws that govern it.

Comparing this situation to a car that functions without an engine, we perceive the absurdity of this logic. Just as a car needs an engine to function, the universe needs a creator to exist. God and the laws of physics are not in conflict, as some suggest. On the contrary, the Scriptures affirm that God created these laws.

The big bang theory becomes problematic for creationist faith when it eliminates the idea of an absolute beginning, from nothing; it considers creation from particles and energy concentrated at a singular point. By considering only particles as a starting point, it opens space for disbelief, contradicting what the Scriptures affirm: that God created the universe from nothing. So, it is important to recognize that even those who do not accept God as the creator but recognize that the universe arose from nothing, face similar dilemmas.

The logical explanation for the creation of the universe, as presented by the Scriptures, reveals a meticulously organized and planned process by God from nothing. Each element of the universe was created with a coherent logic, representing a whole planning and divine desire. Meanwhile, the theory of spontaneous creation, advocated by Stephen Hawking and other proponents, argues that the universe arose from nothing, propelled by the laws of physics, such as gravity.

Although the theory of spontaneous creation may be considered intellectually respectable, it lacks coherent logic. While a machine that makes it rain hamburgers would be dismissed as absurd, the idea that the universe arose from nothing through the laws of physics is widely accepted in academic circles. This contradiction challenges human understanding of the origin of the universe and life.

Faced with this dilemma, humanity is confronted with the anguish of accepting or rejecting the idea of a divine creator. While some theories challenge the traditional logic of creation, it is important to recognize that the truth about the origin of the universe may be beyond human comprehension. The search for the true origin may lead to surprising discoveries, but it can also generate more questions than answers.

The biblical narrative affirms that the creation of the universe was the result of the word of God, as expressed in passages like "Let there be Light, let there be Sky, let there be Earth." This divine view is reinforced by Ps 133:6, 9, which highlight God's power over the stars and galaxies, calling them by name and commanding the entire universe with his voice.

The idea of a multiverse, which suggests the existence of countless galaxies beyond our own, only reinforces the power and magnitude of

God, who calls each of them by name and creates them out of nothing. This understanding underscores the infinitude of divine power, which transcends human comprehension and inspires admiration and praise for all creation.

Furthermore, there are intriguing accounts from ministers and scientists exploring the connection between the music of the stars and worship of God. These discoveries demonstrate how all creation, from the stars to the smallest particles, echoes praises to the Creator, as expressed in the book of Psalms.

The study conducted by a wise Japanese revealed that each cell of the human body emits a musical note, demonstrating the complexity and harmony of divine creation. This leads us to consider what was the motivation behind the creation of the cosmos. The Scriptures offer only subjective clues about the motives of the Creator, suggesting a fervent desire for expression and revelation of his own nature.

God was not compelled to create, nor did he face any crisis or despair. His decision to create was an act of free will, a fervent desire to become known and share his essence with his creation. In creating the universe, God demonstrated his freedom and his inherent goodness, revealing himself as love and sharing his loving nature with all creation.

John 3:16 emphasizes this divine love, showing that God so loved the world that he gave his only Son so that whoever believes in him shall not perish but have eternal life. This is the essence of creation: an act of love and kindness, an expression of the divine desire to share his own nature with his creation.

Therefore, as we contemplate the vastness of the cosmos and the beauty of creation, we are invited to recognize the love and goodness of God that permeate all things. Every star, every cell of our body, resonates with praise and worship to the Creator, bearing witness to his greatness and majesty. Thus, we conclude this discussion about the creation of the cosmos, marveling at the beauty and purpose of all creation, and inspired to glorify and praise the name of the Lord. Hallelujah!

# Chapter IV

## The Creation of the Universe by Fiat

FIAT IN LATIN MEANS "let it be done" or "may it be done." It is a word often associated with the expression *fiat lux*, which means "let there be light," a phrase used in the biblical narrative of creation, in the book of Genesis. We delve into a profound investigation of the concept of creation by fiat. The fiat, in this context, has nothing to do with the car from the popular Italian automotive company, but rather the creative act through light. It is the word of light that gives birth to the cosmos.

As we contemplate the singularity of this act, we encounter the wonder of the plurality that emerges from divine creation. The cosmos, in its diversity, reflects the grandeur of the Creator. However, we cannot ignore the challenges faced by the early chapters of Genesis. Criticisms and attacks, especially from scientific and atheist groups, attempt to discredit the narrative of creation. Nevertheless, instead of a conflict between science and faith, we find an opportunity for fruitful conversation. The interpretation of Scriptures can enlighten our scientific understanding of the universe, and vice versa.

In this context, reflection on the progressive process of creation, as described in the Scriptures, arises. The journey of creation was not instantaneous but rather a careful and deliberate process. The organization of divine creation is another fascinating aspect to explore. Light, in its symbolic essence, plays a crucial role in the structuring of the cosmos, revealing the order established by the Creator. Therefore, this chapter invites us to a search for knowledge and deepening in the understanding of Scriptures and the Christian faith. It encourages us to face challenges in interpreting the Scriptures, thus strengthening our faith and our understanding of the divine.

Christianity should yearn for knowledge and understanding of the Scriptures. We cannot be content with the words of preachers without examining the Scriptures themselves, as the noble Bereans did. The quantity of videos or popularity on the internet does not guarantee knowledge of the Scriptures; often, we find only superficiality and lack of theological foundation.

Just as American art evolved when people began to delve into and understand all the innovative proposals for the development of art, we also need to seek a deeper understanding of the Scriptures. We must reject what we previously considered useful but now realize lacks solid foundation.

Depth in the knowledge of Scriptures is essential for the church. Jesus warned the Pharisees and Sadducees for their ignorance of the Scriptures, showing that the true power of God is linked to the knowledge and understanding of his Word.

The revelation of God's power occurs through his Word, and this revelation is especially evident in the theme of creation. It is through the Scriptures that we understand the greatness and sovereignty of God, and there is no other theme that reveals the divine power so clearly.

Therefore, it is imperative that we seek to know the Scriptures more deeply. Lack of knowledge is a cause of destruction, as warned in Hos 4:6. God rejects those who reject knowledge, so we must strive to grow in the grace and knowledge of our Lord and Savior Jesus Christ.

The desire to learn more should be constant in our hearts, for we can only teach when we are willing to learn. Continuous learning is essential for spiritual growth and effective testimony of biblical Christianity.

Christianity must have a fervent desire to learn and know the Scriptures. We should seek the depth of fundamental doctrines, including creation, which is the foundation of all theology. We cannot be satisfied with superficialities, for the understanding of these themes shapes our faith and our ability to teach and defend the truth.

Since 2011, I have been lecturing and writing about the early chapters of Genesis, and I have understood the vital importance of them for our faith. These chapters are not just a historical narrative; they are the basis of our understanding of God, the universe, and humanity.

Interpreting them correctly is essential for building a solid theology and avoiding heresies.

It is sad to observe how many Christians have lost faith due to lack of knowledge of the Scriptures, especially the initial chapters of Genesis. Many have been confronted with philosophical and scientific ideas that challenged their faith, and without a solid understanding of the Scriptures, they ended up drifting away. Therefore, it is urgent that we awaken to the need to study and deeply understand the Scriptures. We cannot neglect any detail, for every word is inspired by God and is essential for our faith. When addressing the theme of creation by fiat, we must understand the meaning of this term. The fiat represents God's creative command, his word that brought into existence all that we see and know. Even before there was anyone for God to speak to, he was in absolute control of all creation.

In the beginning, God spoke to himself, as narrated in the Scriptures, swearing to Abraham by his own divinity. In this act of swearing, he revealed the greatness of his sovereignty over all things, promising blessings that only he, the Creator of heaven and earth, could offer (Gen 14:22).

In creating the universe, God did not address an external entity but himself. He called into existence that which did not yet exist, demonstrating his absolute power over nothingness. This act of creation by "fiat" reveals not only God's creative ability but also his sovereign control over all creation (Heb 11:3).

Even before the creation of the world, God had already chosen those who would be in it. This election is not based on human merits but on God's sovereign will, which planned salvation even before the world existed. Jesus' sacrifice on the cross was the fulfillment of this divine plan, revealing God's eternal love for his creation (Eph 1:4).

The power of God's word goes beyond mere communication; it is the instrument through which he creates and sustains all existence. Just as Jesus calmed the wind and the sea with a word, God continues to exert his control over creation, manifesting his power and sovereignty (Mark 4:39).

In the biblical account, when the disciples faced a storm at sea and cried out to Jesus, he rebuked them for their lack of faith. This narrative illustrates the absolute confidence we should have in divine control over our lives. Since creation, God has not only brought the cosmos into existence but has also exercised dominion over it (Matt 8:23–27).

True faith recognizes that every aspect of our lives is under the care and command of God. His control is not limited only to the initial creation but encompasses every detail, every particle of our existence. This profound understanding of divine sovereignty elevates faith to an unparalleled level (Ps 135:6).

The concept of "fiat" in creation reveals not only the act of creation but also God's continuous control over his work. He not only created but also continues to govern and direct all that exists. This understanding challenges limited conceptions of faith and leads us to recognize God's absolute authority over the cosmos and our lives (Gen 1:3).

In Scripture, we find numerous references to divine sovereignty, demonstrating that God not only created but also rules over his creation. He is the Creator and sustainer of all things, directing what does not exist as if it did. This fundamental truth challenges us to trust entirely in God, the only one who rules and controls all things, from the beginning of time (Deut 32:6; Job 4:17).

God's sovereignty transcends our human understanding, reaching even what does not exist. He not only creates out of nothing but also rules and controls everything that exists and will come into existence. His command is powerful and effective, bringing into existence what he desires through his sovereign voice (Gen 1:3).

God is not limited to what already exists but has authority over nothingness. He directs and governs over emptiness, turning it into reality by the word of his mouth. This is the God of Scripture, whose power extends beyond the boundaries of the visible creation, encompassing also the invisible and the unimaginable (Ps 33:9).

Scripture invites us to worship this God who rules over what does not exist, recognizing his absolute authority over all creation. He is the Creator of heaven and earth, our help in times of need. No limitation can be imposed on him, for he brings into existence only what he can rule and control (Ps 95:6; 121:2).

For those who oppose faith, confronting the truth of divine sovereignty over nothingness can be disturbing. Many who considered themselves knowledgeable about faith have been challenged by this deeper understanding of Scripture, confronting them with their own shallowness. Nevertheless, the truth remains: God is absolute in his power and control over all things (Isa 46:10).

Anything that God creates reflects his own greatness and dimension, for nothing can be greater than him. He shapes and directs creation according to his sovereign will, leaving no room for chaos or confusion. The myth of a lost world, abandoned to the hands of the devil, must be dispelled, for God rules and controls all things according to his perfect will (Isa 45:7).

It is essential to understand that God is not limited in his power or control. He is absolute, omnipotent, omniscient, and omnipresent. Nothing is beyond his authority, and nothing can challenge his sovereignty (Jer 32:17). Therefore, discard misconceptions that evil has dominion over creation. Though evil exists as a conceptual reality, God is the source of all goodness and love (Jas 1:17).

Evil entered the world through Satan's rebellion, not as a divine creation, but as a distortion of the original perfection. God is inherently good and compassionate, radiating love for all his creation. His eyes are constantly turned towards us with love and tenderness, for we are the object of his eternal affection (Zech 2:8).

Therefore, understand that God is the guardian of his creation, not allowing evil to prevail over his redemptive purpose. He is the consuming fire that purifies and the hammer that smashes the strongholds of sin. Trust in God's unwavering love for you and know that he is in control, directing all things towards the fulfillment of his eternal plan (Rom 8:28).

God trusts you more than you imagine. He is by your side every day, motivating and encouraging you, even when you feel discouraged. However, you may not be fully grasping his message of encouragement; perhaps you are listening to other voices than those of the Scriptures.

It is essential to recognize that God is in control of your life and that he has the best plans reserved for you since eternity. Therefore, do not allow the difficulties and challenges of life to obscure the confidence you should have in God's power and goodness.

Creation by fiat demonstrates God's sovereignty over all things. He creates and directs according to his sovereign will, and everything he does reflects his greatness and perfection. Therefore, discard misconceptions that the world is abandoned to chaos or the control of evil. God is absolute, all-powerful, and compassionate.

Understand that you are not just a physical body but possess a spiritual essence that reflects God's creative work. He designed and planned you from the beginning, and you are part of his eternal plans. Although we may not fully understand God's works, the Scriptures provide us with guidelines to live in accordance with his will. Trust in God's wisdom and love, recognizing that he is the sovereign potter who molds and leads our lives according to his perfect plan (Isa 55:8–9).

Therefore, surrender to trust in God's voice, setting aside all anxiety and uncertainty. He is your greatest defender and guide, and his love for you is eternal and unwavering. God created each of us according to his divine desires and plans. He shaped our lives uniquely, and if necessary, he remodels them to fulfill his purposes. Our lives are entirely in God's hands, and he exercises absolute control over every aspect of them.

The Scriptures provide clear guidelines on God's role in the creation and governance of our lives. We must recognize his sovereignty and trust in his love and care for us. It is important to understand that for God, nothing is truly empty. He brings all things into existence from nothing, and to him, nothing is an inexhaustible source of possibilities. He simply speaks to what does not exist and brings into existence what he desires. Therefore, when Jesus instructed his disciples to go out without taking anything with them, he had already provided everything they needed. God is able to supply all our needs, for he is the Creator who controls and governs everything he has brought into existence.

It is crucial to understand this idea of creation by fiat, as it demonstrates the power and sovereignty of God over all creation. He continues to exercise his command and control over everything he has created, and we can fully trust in his wisdom and goodness. Therefore, let us set aside worries and anxieties, trusting in God's providence and care for us. He is the sovereign potter who molds and leads our lives according to his eternal and perfect plans.

Through divine command, God brought into existence everything we know, including the heavens and the earth. The concept of *fiat*, a Latin expression for command, leads us to understand that from the beginning, God was in absolute control of his creation and continues to rule over everything he created.

Jesus, in his teachings, reinforced the idea that nothing is outside of God's control. Not even a hair falls from our heads without divine consent. Everything is under God's dominion, and he directs all things for his glory and praise.

When we refer to the heavens and the earth, we are speaking not only of the cosmos and the planet we inhabit but also of the place where God resides and where the saved will find their eternal dwelling. According to Scripture, the heavens will be renewed and joined with the earth, creating an eternal paradise.

Considering the vastness of the cosmos, with over two hundred billion galaxies, we may wonder if it would be too much for God to give us a personal galaxy. However, in light of Scripture, an individual galaxy may seem insignificant in the perspective of eternity, where there will be no limitations of time and space.

Human desires and longings, expressed even in cartoons, reveal the aspiration for grandeur and expansion. God understands the desires of the human heart and, in his sovereignty, will lead us to the fullness of his presence, where we will find complete satisfaction and fulfillment.

The concept of divine creation leads us to reflect on the grandeur of the cosmos and the extent of God's power. While humans express ambitions and desires for conquest, like Napoleon, who aimed to dominate the world, God reveals his infinite power by creating and ruling over all that exists.

Even in the face of the vastness of the cosmos, God is beyond the limitations of time and space. He dwells in the heavens of heavens and, with a simple blink of an eye, can be anywhere. Jesus exemplifies this divine power, demonstrating that faith can move mountains and even stop the Sun.

However, often we limit our understanding of divine power by imagining Jesus as a fragile being nailed to a cross. We forget that God is above all things and that his greatness transcends our understanding.

Although science presents the cosmos as infinite, we must remember that God created all of this with a purpose. He is our Creator, and we are part of a much larger plan than we can imagine. We are not the center of the universe, but we are part of a creation loved by God.

Who is this that God has in mind when creating? Who is it that, as Father, Son, and Holy Spirit, decides to create in his likeness, breathing the breath of life so that they may be like Jesus, his Son, expressing his image and likeness? What could fill their hearts with envy, if not the perception that they will be creatures with authority and power to rule?

Let's think about this for a moment. Lucifer, once the ultimate authority, felt envy of this new project of God. It is not uncommon to feel envy of those who seemingly have more power, more authority, more governance. And who would be Lucifer's competitor if not human beings, made in the image and likeness of God, destined to rule over all creation?

God did not create us just to express his love. From the beginning, when he created the cosmos, he loved us. The biblical account of creation is not just artistic poetry but a historical account of events that occurred. We are part of a grand plan, where every detail reflects the greatness and majesty of our Creator.

The big bang theory, although a valid scientific explanation, is just one of the many ways to understand creation. The light mentioned in the third verse of Genesis represents not only the creation of celestial bodies but also the organization and purpose that God bestowed upon all creation.

It is important to recognize that science is welcome in our investigations, but its findings are still limited in the face of the depth of the Word of God. The Bible contains a wealth of knowledge that far surpasses our current understanding.

Therefore, as we continue our journey of discovery and learning, we must remember that we are just children beginning to grasp the vastness of God's knowledge. Each new scientific theory, each new discovery, only leads us to a deeper understanding of the greatness and complexity of divine creation.

As we delve into our studies, we realize the immensity of the knowledge we can still acquire. From particle accelerators to the mysteries of dark matter that permeate the cosmos, there is so much to be explored, so much to be understood. My heart longs for a deeper advancement of science in this direction, for more theories and discoveries that can help us grasp the complexity of creation.

Science, with its theories and experiments, brings valuable insight to our understanding. It helps us discern the mysteries of the cosmos, unravel the enigmas of nature. But even so, its findings are still limited in the face of the breadth of the Holy Scriptures.

Each new theory, each new discovery, brings us a little closer to understanding the divine plan. However, it is important to remember that the Scriptures contain a wisdom that transcends any human knowledge.

The creation account in Genesis is more than just artistic poetry; it is a historical account of the events that occurred at the beginning of time. Even though the verses may not fit together harmoniously, each word, each phrase, reveals the greatness of the Creator and his work.

God, in his uniqueness, created everything that exists. He hovered over the chaos, bringing order and beauty to the universe. And while we may continue to explore the mysteries of creation, we must always remember that our knowledge is limited in the face of the grandeur of the divine plan.

May we continue our journey of discovery, seeking both in science and faith a deeper understanding of the wonder of creation. For it is through this understanding that we can contemplate the greatness and majesty of our God.

I understand the importance of clarifying the issue of chaos present in the second verse of Genesis after the creation account in its first verse. This apparent contradiction leads us to a profound reflection on the nature of language and how we try to understand and describe divine events.

The word "chaos" in this context should not be interpreted in the strict sense of absolute disorder, but rather as a description of the initial condition of the Earth before being fully organized by God. The Hebrew term used can be understood as a way of expressing the undifferentiated and unformed condition of the cosmos before God's creative intervention.

This apparent incongruity challenges us to recognize the limitation of human language in attempting to describe the mysteries of divine creation. The concept of anthropomorphism, where we attribute human characteristics to God in an attempt to understand him, is also relevant here. Scriptures often use anthropomorphic language to communicate spiritual truths in a way we can grasp, even if it doesn't fully capture the divine reality.

Furthermore, the discussion about the limitations of ancient languages, such as Hebrew and Greek, highlights the complexity of translation and interpretation of Scriptures. Scholars face challenges in accurately conveying the original meanings of sacred texts into modern languages.

Therefore, as we address this theological issue, we must acknowledge the depth of divine language and the humility required as we attempt to understand the mysteries of God. It is through the constant pursuit of truth and the guidance of the Holy Spirit that we can draw closer to a clearer understanding of God's will and purpose revealed in Scriptures.

It's not mere verbosity, but genuine reverence that motivates my approach to teaching Scriptures. If I am to dedicate myself to study or preaching, let it be with reverence, avoiding slips and heresies. I could not bear to see my voice propagate distorted interpretations of Jesus, the Source of Living Water.

## CHAPTER 2: THEOLOGICAL COMMITMENT AND DEDICATION TO STUDY

From the early steps in theology, I faced criticism for my conservative approach. I abandoned my previous career to immerse myself in study, convinced it was a necessary calling. While some opt for medicine or psychology, I chose theology, aware of the seriousness of this commitment.

Superficiality in theological study is an affront to the seriousness of faith. I refuse to tolerate theological absurdities that compromise the credibility of Christianity. The quality of teaching must be preserved, for what is taught shapes people's faith.

I recognize my responsibility to lead people with wisdom and discernment. It is essential that the Scriptures be taken seriously, without concessions to distorted interpretation or superficial theology. Any deviation undermines the integrity of the Christian faith.

Jesus taught that we should have compassion and mercy, as we are all prone to error. However, it is essential to maintain a commitment to seriousness in the study and teaching of the Scriptures. True understanding of the Scriptures is only possible when we take their study and application seriously.

Amidst what appears to be chaos, arises the reflection on the presence of an underlying order, even in chaos theory. From the human perspective, situations may seem disordered, but to God, there is an implicit order, revealing itself in every aspect of creation.

Analogies such as a natural forest compared to a eucalyptus plantation highlight how our perception can be limited. Although it may seem chaotic to our eyes, God operates with order and purpose, even in what appears confusing to us.

Belief in chance is a manifestation of our ignorance, especially regarding God's will and control. Trusting in a God whose will is not perfect for our lives would be a contradiction. The concept of chance is challenged by the notion of a sovereign and provident God.

Understanding creation as an act of divine command reflects the absence of probability or chance. God's will is the foundation of order, ensuring that every event serves a specific purpose in his sovereign plan.

For those who trust in God, life is not governed by chance but by a perfect divine will. Recognizing divine order allows us to face challenges with faith, knowing that every aspect of life is under God's sovereign care.

The idea of chance is challenged by the ability to correct errors and calculate future actions. For those who believe in divine providence, every step is guided by God's sovereign hand, allowing us to learn from mistakes and move towards his perfect will.

Examining the creation narrative in Genesis reminds us that every historical event is imbued with purpose and meaning. There is no room for chance in the theological worldview, only the certainty that God's perfect will is fulfilled in all things.

Reflection on creation and human history leads us to question: is the creation account a creed, a poem, or a historical event? Any serious theologian of Scriptures will acknowledge its historicity. Creation inaugurated time and everything we know.

Although we may imagine the existence of chance, it is important to remember that for God, there is no chance in the cosmos. Einstein's assertion that "God does not play dice with the universe" emphasizes that all events can be foreseen by him.[1] God knows all possibilities and knows the outcome of each situation.

If God is sovereign over all things, then he also rules over human history. Each individual has a place and time in history, determined by divine will. Our existence is part of a perfect plan laid out by God from the beginning.

God began writing our story from the end, ensuring that every event fulfills his purpose. Our human responsibility coexists with divine providence. Even if we try to change or plan our future, God's will prevails. Our story is already written, and he will complete it according to his plan.

The truth of providential history frees us to worship, pray, and obey Scriptures, knowing that God has chosen every aspect of our lives. We can make decisions, strive, and plan, but nothing will change what God has planned for us. Our story is complete in his hands, and we can find peace in trusting his perfect plan.

Throughout history, great thinkers in physics and astrophysics refer to chapters 1, 2, and 3 of Genesis, which deal with the creation of the known cosmos, because within what they study, when they develop a new theory or thought, it always comes close to what they end up discovering and concluding.

---

1. Einstein, *Born-Einstein Letters.* This source contains the correspondence between Albert Einstein and Max Born, which provides insights into Einstein's thoughts and his often-quoted phrase, "God does not play dice with the universe."

The continuous exploration of the first chapters of Genesis reveals a fascinating intersection between physics and Judeo-Christian faith. Renowned thinkers in physics and astrophysics should reconsider these texts, recognizing in them a resonance with their emerging discoveries and theories. This intrinsic connection between biblical cosmology and contemporary scientific exploration highlights the importance of an enriching dialogue between both fields of knowledge.

Especially concerning these three verses that we are continually studying every week. Many thinkers, a significant portion of them, have some understanding of theology. They are children of Christians or Jews, many of whom attended church groups when they were young. However, they encounter conflicts regarding their faith due to a lack of understanding of this topic in the Scriptures. These individuals end up abandoning their walk with God due to instructional errors.

This is a crucial point to consider. Many individuals, including those with theological training and deep immersion in the Christian faith, may find themselves at a crossroads when confronted with intellectual and scientific challenges. The absence of a solid understanding of God's word can lead to confusion and ultimately to the loss of faith. It is imperative that Christians and Jews confront these issues head-on, providing robust theological education and encouraging a continuous pursuit of scientific knowledge.

The issue of atheism and its dilemmas against God has been refuted by me since 1997, and today I have many atheist friends. However, many have also abandoned atheism to serve God. Everything I have been doing for a long time is delivering this message from Genesis chapter 1:1-2, which provides a solution against the doubts and loneliness of men.

There is an urgent need for a return to deep study of the Scriptures. The commitment to explore the early chapters of Genesis, from its descriptions of creation to its broader theological implications, is a mission I have embraced for years. I have witnessed many who, once distant from the church, are now seeking reconnection, eager to understand more deeply the biblical truths and how they relate to contemporary scientific discoveries.

Scientific theories about the origin of the universe do not exclude the possibility of the existence of a divine Creator. On the contrary, they often raise questions that transcend purely scientific scope and point to the need for a broader and more holistic understanding of existence.

The atheist has the right to claim that there is no evidence for the existence of God, but this depends on what evidence will be accepted by the atheist. For example, for me, the greatest evidence of God's existence is Jesus resurrected from the dead. Jesus is the incarnate God who became man. I maintain an open dialogue with atheists, arguing that based on my scientific and theological knowledge, it is impossible to deny the possibility of God's existence. Scientific theories about the creation of the cosmos open up space for reflection on the divine. I emphasize that for Christians, Jesus' incarnation itself is tangible evidence of God's existence. He came as God in human form, walked among us, revealing to us the truth about his divinity.

According to the scientific theory proposed by Father Georges Lemaître, there is a description of a cooling period after the emergence of the universe, when everything was plunged into darkness. This narrative contrasts with the biblical account in Genesis, where we see that God is the source of light itself. It is crucial to highlight this difference, as it helps to avoid confusion when discussing matters of faith with those who are not familiar with Christian doctrine.

As we delve deeper into the study of creation, it is important for us to understand the meaning of the Hebrew verb *bará*, which denotes God's creative activity. He not only created, but also organized and separated, bringing light to distinguish between light and darkness. This initial separation is crucial for our understanding of the divine creative process.

It is necessary to clarify that creation did not involve the manipulation of pre-existing matter, as suggested by some theological interpretations. On the contrary, it was an act of creation *ex nihilo*, out of nothing, as the Scriptures affirm. Creation emerged from the primordial chaos but was brought into order and light by God.

Sometimes we encounter scientific theories that challenge our understanding, such as the idea that the universe arose from the laws of gravity. These theories may seem bewildering, but it is important to remember that our faith is rooted in the authority of the Scriptures, which reveal to us a Creator God who is beyond human comprehension.

We must be careful when using terms like "evolution" improperly. The process of divine creation cannot be equated with biological evolution. Instead, we should refer to it as "spontaneous creation" or "spontaneous generation," acknowledging God's creative power over all things.

The verb *bará* is of paramount importance, as it places God as the absolute Creator. If the biblical text were to begin differently, such as "In the beginning was God," our understanding would be drastically altered. However, by stating that "In the beginning God created the heavens and the earth," our entire perspective shifts, and God is recognized as the primordial agent of creation. This is fundamental to our faith, as it establishes God as the center of existence and the originator of all things.

As we study physics and encounter concepts like the big bang, we are confronted with the need for a primordial agent to explain the beginning of the universe. We cannot go beyond this initial point, and this is where the figure of God, the self-existent and revealing Being, comes in. Although we may not have adequate words to describe his greatness, we recognize his action before the first cause that propelled all creation into existence.

It is fascinating to observe how modern physics reveals the precision and care in the creation of the universe. Concepts like dark energy, with its delicately balanced influence on the expansion of the cosmos, highlight the exactitude with which everything was designed. This meticulous precision leads us to reflect on the wisdom and power of God, whose hand is behind every aspect of the observable universe. Even great minds like Einstein were led to reconsider their ideas in the face of the complexity and harmony of the cosmos.

Advancements in physics, from the days of Lemaître to the most recent discoveries, continue to confirm the idea of an expanding universe. Einstein, upon revising his own theories, acknowledged this reality, which is now widely accepted. However, even with the understanding that the universe is constantly evolving, scientists still face challenges in reconciling their theories with religious narratives like Genesis chapter 1, verse 1.

One attempt to circumvent this difficulty is the theory of the multiverse, which posits the existence of multiple universes beyond our own. While this theory may offer some plausible explanations, such as the variation of physical laws in different universes, it does not eliminate the fundamental question of the origin of the universe. Even if multiverses are infinite, their creation still demands an explanation for their emergence from nothingness.

These complex theories, like the multiverse theory, highlight the impasse faced by science in trying to reconcile its findings with traditional religious narratives. While scientific knowledge is a gift given by God to humanity, it may, at certain points, seem to contradict spiritual truth. The pursuit of knowledge is an essential part of the human journey, but it is important to recognize the limits of human understanding and remain open to the possibility of truths that transcend the realms of science.

Stephen Hawking's speculation that our universe may have arisen from the laws of physics itself causes distress in many young people, especially those who are Christians, but in fact God used these laws to create the universe. The physicist's renown and the nature of his theories have a significant impact on students' minds, leading many to question their beliefs. Therefore, it is crucial that the church is well-grounded in the understanding of Scriptures, especially the book of Genesis.

The book of Genesis, particularly chapter 1, verse 2 and beyond, is often underestimated in Protestant and Evangelical churches. However, a careful analysis of the Hebrew concept of *bará* reveals a connection with the concept of an absolute beginning, similar to that proposed by the big bang. This coherence between Scripture and the scientific theory of the big bang does not pose a problem for Christians, but rather for atheists.

The big bang theory, far from contradicting Scripture, strengthens the Christian faith by showcasing the greatness of the Creator and the created, finite nature of the universe. It is essential to understand that this theory was not devised to refute Scripture, but rather to explore and explain the complexity and origin of the universe. Thus, instead of being a threat to faith, the big bang theory is a confirmation of the biblical view of creation and the transcendent nature of God.

The multiverse theory proposes the existence of multiple universes beyond our own. Although this idea is not necessarily contradictory to the Christian faith, it raises questions about the origin and nature of the universe. Science, as a tool given by God, often challenges our conceptions of truth, which can be disconcerting for some. However, it is important to remember that study and intellectual growth are encouraged by Christianity, as they reflect the pursuit of knowledge that God has granted us.

The multiverse theory, although speculative, challenges traditional conceptions of the universe's creation. This may cause discomfort among young Christians, especially when confronted with the renown of figures like Stephen Hawking, who proposed this idea. In this context, it is crucial that Christians offer guidance and understanding on how to reconcile faith with scientific discoveries, especially regarding the book of Genesis.

The book of Genesis, particularly verses 2 and 3 onward, provides an important basis for understanding the creation of the universe. The

Hebrew concept of *bará* as a new beginning can be compared to the big bang, providing a coherent interpretation between scientific theory and Scripture. By studying these concepts, Christians can strengthen their faith and gain a better understanding of the grandeur of divine creation.

The big bang theory is not necessarily a threat to the Christian faith, but rather to the atheistic perspective that denies the existence of a Creator. Contrary to some beliefs, this theory was proposed by a Christian with the intention of revealing the greatness of God. Therefore, in-depth study of Genesis and other biblical passages related to creation can strengthen the faith of Christians and help them respond to scientific challenges in an informed manner.

The understanding that the universe had a beginning and is not eternal aligns with Scripture and the theory of cosmic expansion. The concept of *bará* as a regeneration challenges us to reinterpret Scripture in light of scientific discoveries, allowing for a deeper understanding of divine creation and strengthening our faith in the existence of a creator God.

When the verb *bará* is deeply analyzed, it highlights an absolute beginning of time and space, forcing a reanalysis of everything that was previously taught. Genesis 1:1 proclaims the grandeur of God as the creator of all celestial things, demonstrating that he is absolute over his creation, even in the face of a hundred billion multiverses.

The multiverse theory does not pose a threat to the Christian faith because even if they exist, all multiverses would be under the dominion of the Creator. This reflects the idea of a God unlimited in power and greatness, as described in Scripture.

The analysis of Gen 1:1 reveals it as an independent declaration of how God began to create, while verse 2 provides an explanation of the initial state of the universe before the creation of light.

The presence of dark matter in the cosmos, mentioned in Scripture, indicates an ancient perception of the density of darkness, which can be interpreted as a reference to the dark matter permeating the universe.

Despite the complexity of the universe and scientific theories, faith is not shaken, as God is seen as the one who is above all things and who can create whatever he desires with his word.

So, the issue of dark matter is addressed, as 75 percent of the cosmos is composed of this substance present throughout its extent. Genesis 1:2 offers an explanation of the initial state of the universe before the creation of light, which is in accordance with the big bang theory, where there were darkness and chaos before the emergence of light and the organization of celestial bodies.

The concept of *bará* leads to the conclusion that there was pre-existing matter when God created the heavens and the earth, indicating the existence of a mass upon which the Spirit of God hovered and shaped, bringing form and life. The Holy Spirit is seen as God's agent of life, acting not only in the lives of believers but also in those of non-believers, preserving life in the cosmos and restraining iniquity to a certain extent.

In addition to the presence of dark matter, which comprises a significant portion of the cosmos, it is interesting to observe how Gen 1:2 offers an intriguing explanation of the initial state of the universe. It describes a scenario of chaos and darkness, where the Earth was formless and empty, even before the emergence of light. This description surprisingly aligns with the concepts of the big bang theory, where the universe undergoes stages of cooling and darkness before organizing and light emerges.

The analysis of the *bará* concept reveals an interesting approach to creation. It is not just a creation out of nothing, but rather an act of God's free will, demonstrating his unlimited power. In this sense, the Spirit of God is presented as the agent of life, not only intervening in the lives of believers but also in all living beings, maintaining order and preserving existence in the cosmos.

When it comes to the theme of salvation, we observe an interesting duality in God's perspective. On one hand, he demonstrates unconditional love for the cosmos, offering the opportunity for eternal life to all. On the other hand, he also manifests wrath against sin and the sinner who rejects the message of salvation. This duality reflects the complexity of the relationship between God and his creation, highlighting his justice and infinite love.

By exploring these concepts, we can perceive how the Scriptures offer a deep and comprehensive view of creation, life, and salvation. Every detail reveals the greatness and majesty of God, inviting us to contemplate his work and seek a deeper relationship with him.

Reflection on God's weariness, as mentioned in the text, leads us to understand that it is not a physical effort on God's part, but rather figurative language to highlight the importance of rest for humanity. God, in his infinite wisdom, established the Sabbath as a time of rest and renewal, not because he himself needed rest, but so that his children would learn to value and preserve their own physical and mental health.

The idea of *bará* leads us to understand that creation was not a conditioned or obligatory act for God, but rather a spontaneous expression of his creative will. He is the Creator par excellence, and his essence is creativity itself. Every aspect of the cosmos reflects his glory and power, inviting us to contemplate his greatness through the work of creation.

Nature itself becomes a silent witness to the power and greatness of God, like a gospel of creation that constantly echoes in our hearts. Even in the face of scientific theories that seek to explain the origin of the universe, the greatness and complexity of creation only magnify the image of God.

Therefore, faith is not based solely on the Scriptures, but also on contemplating nature and understanding the divine work that sustains all existence. Through faith, we can see beyond theories and discover the true source of power and sustenance found in the word and work of God.

Modern physics, for example, reveals to us a cosmos governed by precise and constant laws that sustain the existence and harmonious functioning of all matter and energy. The big bang theory, widely accepted by the scientific community, describes the origin of the universe from a primordial singularity, an event of cosmic expansion that gave rise to space, time, and all matter as we know it. This theory not only finds interesting parallels with the biblical narrative of creation but also helps us understand the vastness and complexity of the cosmos.

On the other hand, the idea of multiverses, while still a subject of debate and speculation, represents an intriguing attempt to understand the nature of the universe and its possible origin. The suggestion that our universe may be just one among countless other parallel universes raises profound questions about the nature of reality and the existence of life forms beyond our comprehension.

However, regardless of the scientific theories seeking to explain the origin and nature of the universe, the grandeur and complexity of creation continue to point to a transcendent Creator whose power and wisdom are beyond our full understanding. Science and faith can therefore coexist and even complement each other, offering distinct perspectives that enrich our understanding of the world around us.

The complexity and precision of the universe point to an order and purpose that go beyond chance. The idea that everything emerged from nothing, governed only by chance, is hard to conceive in the face of the evidence of the precision with which the universe operates. The laws of physics, which keep everything in place and govern the functioning of the cosmos, are seen as manifestations of the word of God.

These laws are not mere abstractions but rather the expression of divine will, released by the word of God to sustain and govern all creation. Even in the face of scientific theories about the origin and operation of the universe, faith affirms that it is the word of God that keeps everything in order.

The precision with which the universe is organized, from the positions of the planets to the regulation of fundamental forces, is seen as evidence of divine action. Even events that seem fortuitous, such as the deviation of a meteor by mere centimeters, are seen as part of the divine plan, showing God's constant and precise action in the creation and sustenance of the cosmos.

In light of this, the idea that everything is governed by chance is rejected in favor of belief in a God who rules and guides all things. Life is not left to chance but is guided by the Spirit of God, who is recognized as the true ruler of the universe and the guide of each individual's life. Thus, understanding the cosmos as a result of the word of God brings a perspective of order, purpose, and meaning to everything that exists.

The concept that everything emerged from nothing is an irrefutable truth in modern times, something that was once hard for many, even in physics, to accept. The idea that something can arise from nothing challenged human understanding, which always associated existence with the presence of pre-existing matter. However, the Scriptures consistently affirm that God is the creator, capable of bringing into existence everything that did not exist before.

This idea applies to the verb *bará*, which also relates to creation from nothing. Jesus, as the Word of God, acted as the creative agent, as recorded in the Gospel of John. He is the ultimate expression of the fullness of God, representing the Father on Earth. Often, the conception of God is distorted, whether through images of an old man sitting on a throne or through interpretations based on extraterrestrials.

However, the true understanding of God is found in Jesus, through an intimate relationship with Him. Jesus himself claimed to be the way to the Father, and it is through him that we can truly approach and know God. He is not just an intermediary but the very manifestation of the divine on Earth.

The exegesis of these verses, though challenging, finds support in various parts of Scripture, reinforcing the idea that God created the cosmos from nothing. It is this faith in God's creative action that allows us to understand the origin and purpose of existence, even in the face of the mystery surrounding the foundations of the universe.

Creation was not made to remain empty but to be inhabited, as stated in Isa 45:18. Every time the phrase "created the heavens and the earth" is mentioned, it refers to the idea of *ex nihilo* creation, creation from nothing, by the word of God. Imagine if you spoke a word and a car appeared out of nowhere. What would we interpret? That you created from what? From matter, or released a word that was nothing and became something?

That's how God created, by the word of his mouth. This truth, once challenging, has been confirmed not only by Scripture but also by modern science. Physics has evolved to affirm that the cosmos also emerged from nothing, something Christianity has been teaching for over two thousand years. Stephen Hawking even talks about spontaneous creation, but even he admits that the laws of physics also emerged from nothing.

Thanks to science, we can confirm the truth of Scripture without fear. Both faith and science point to the same conclusion: the universe emerged from nothing, by the word of God. If we choose to believe in this truth, we will encounter the existence of God. And even for those who choose to believe that the universe emerged from nothing, it only reinforces the idea that all truths come from God.

It is fascinating how science and faith can complement each other. I myself have always been passionate about physics and mathematics. These subjects lead us to unravel the mysteries of the cosmos, revealing the greatness and complexity of God's creation.

I love the theme of creation, and also teaching about the financial market, it's a passion of mine, after all, who doesn't like making money? I have also been helping people increase their participation in the capital market, which makes me happy in life. On the other hand, I have this faithful commitment to the faith I received from the saints.

Now, about scientific discoveries, it is crucial to teach your sons and daughters that they are creatures. When I say "creatures," I refer to those who struggle with the gospel, like you, the reader, for example. The greatest truth of the verses corroborates that there was a creation. The term "evolution" should not be used in this cosmic context; we should not believe or live in an empty existence and loneliness because God exists. It is more appropriate to speak of spontaneous generation or creation. The idea of evolution implies an emergence without God, while Christianity believes that everything arose from a divine cause.

The Scriptures do not exist to prove the existence of God, but rather to reveal his existence to the world. We study the Scriptures to know God more, not to debate his existence. What the Scriptures make clear is that there was a creative will, a higher intelligence behind everything. Sometimes, it is helpful to use this term to dispel prejudices regarding God.

People often conflict when they lack faith in God or do not properly know the Scriptures. They question how a loving God can order the destruction of infants or send people to hell. But it is important to understand that the Scriptures reveal God's love in a unique way, and this requires a deep understanding of faith and divine revelation.

It's true, there are different interpretations about God's justice and love. For example, when God opened the Earth and killed rebels in the Old Testament, it depends on the individual's conception of death and what happens after it. Perhaps God was saving those people, just as Paul handed someone over to Satan for salvation.

Regarding the issue of hell, it depends on each person's understanding. The Scriptures describe hell as a place of torment, but often this language is symbolic. The worst hell might be the awareness of missing the opportunity to accept God's love message. It is a desperate and agonizing suffering.

Interpreting the Scriptures is essential, and there are various ways to do it. The testimony of the Scriptures is that God created the heavens and the earth when nothing else existed but Him. Verse 2 describes the Earth as void and formless, but this does not mean absolute emptiness or uselessness. It was part of God's creation process, and the Spirit of God was acting to transform chaos into cosmos.

The idea of "hovering" in Hebrew can also be compared to a hen's action when sitting on eggs for twenty-one days, hatching them. In that sense, the Holy Spirit was generating life, just as a hen generates life by incubating eggs. When the Holy Spirit hovered over the Earth, it was bringing life, generating it through the word of God. It is important to understand that life is sustained by the Spirit of God, which remains active within us. It holds back total depravity of sin in the human heart, preventing corruption and destruction from spreading completely. If it were not for the action of the Holy Spirit, humanity would be in an even worse state, with devastating consequences like the destruction of forests and the extinction of various species.

The Holy Spirit acts as a continuous agent of life preservation, even in the face of human depravity and wickedness. It curbs man's desire to destroy, kill, and steal until the judgment of God reaches those people. Human beings, in their sinful nature, are devastators and destroyers, but the Holy Spirit continues his work of preserving life, demonstrating the grace of God that extends to all humanity.

God's divine intervention in history, such as the flood in Noah's days, demonstrates God's compassion in the face of human wickedness and violence. God does not desire destruction, but sometimes intervention is necessary to reduce suffering and wickedness in the world. Schopenhauer offers interesting insights into human madness and suicide, suggesting that, in some cases, they are extreme manifestations of suffering and hopelessness. However, knowledge of Jesus' love and understanding of God's grace can offer a perspective of hope and restoration to those who are suffering.

The attitude of God towards wicked nations, such as the Islamic state in the present day, can be compared to human justice in the face of extreme crimes. Although it is difficult to comprehend God's methods, divine intervention is often necessary to prevent evil from spreading further. God's judgment can be seen as a protective measure against malignancy and wickedness that threaten life and human dignity. Therefore, it is important to recognize that God's goodness may manifest in ways that may seem severe, but aim at preserving life and restoring justice.

Furthermore, when considering the atrocities committed by ancient peoples, such as the Assyrians, it is crucial to understand the extent of human cruelty when there are no moral boundaries. Historical accounts reveal inhumane practices of torture and violence that surpass any modern understanding. God's care in intervening in extreme situations is a demonstration of his mercy and justice, protecting those who are defenseless and suffering at the hands of the wicked.

On the other hand, it is essential to recognize that understanding divine actions often surpasses human judgment capabilities. Although we may question the justice of certain events, such as the flood or the destruction of wicked nations, it is important to trust in God's wisdom and purpose. His comprehensive view and infinite compassion exceed our limited understanding, and his intervention always aims at the greater good, even if we cannot fully comprehend his methods.

Lastly, when observing the role of the Holy Spirit in preserving life and divine intervention, it is necessary to understand that God's grace extends to all human beings, regardless of their condition or behavior. Even in the face of human depravity and wickedness, the Holy Spirit continues to operate, offering opportunities for repentance and transformation. His active presence in history demonstrates God's unconditional love for humanity and his constant pursuit of restoration and redemption of all things.

It is crucial to understand the historical and cultural context when analyzing the Scriptures, especially concerning divine orders for Israel to deal with wicked nations. Understanding the atrocities committed by these peoples reveals the extreme human depravity and the need for divine intervention to protect the innocent and contain evil. Therefore, before passing any judgment on God's actions, it is crucial to consider the complex scenario in which these events occurred and recognize divine justice and mercy in action.

Finally, when contemplating the creation of the universe and everything in it, it is essential to understand that the declaration of "it was good" does not indicate a late discovery on God's part but rather the fulfillment of his perfect desire. Each step of the creative process was planned and executed according to divine will, and the affirmation that "it was good" reflects God's satisfaction in seeing his work complete. Therefore, in studying the Scriptures, it is important to recognize the profound meaning behind these statements and appreciate the perfection of God's plan for creation and humanity.

Throughout the centuries, humanity has contemplated the mysteries of creation and the universe. Divine wisdom and advances in science lead us to reflect on the possibility of a primordial universe preceding ours, mentioned in the Book of Genesis. We may conjecture that verse 1:1 deals with the creation of this previous universe, whose existence precedes the narrative of Earth in Gen 1:2 onwards.

This theory suggests that the ancient universe, with its billions of years of existence, eventually reached a state of stellar death, reflecting the fate awaiting our own universe in the distant future. When God transposed this ancient universe into the new one, he found it plunged into total darkness over the abyss. This divine act of exporting his previous creation into the depths of the abyss placed all created elements in a primordial state of chaos, explaining possible catastrophes both on Earth and on other planets.

After this process of recreation, in which stars and planets underwent a transformation, God established order and called light into existence, giving rise to the universe as we know it. This conception suggests the existence of a prior vacuum, interacting with a new vacuum, whose laws and characteristics are governed by a set of rules different from those we know.

The citation from the Psalms, "Deep calls to deep in the roar of your waterfalls; all your waves and breakers have swept over me," can be interpreted in light of this theory as a poetic allusion to the interaction between universes and the influence of cosmic forces that transcend our understanding.

This conjecture also finds resonance in the hypothesis that the expansion of the universe may be influenced by the gravity of the primordial universe, suggesting an infinite cycle of creation and recreation.

As we consider the possibility of a previous universe, we can reflect on the nature of time and space even before the existence of the cosmos we know. This reflection leads us to a deeper understanding of God's transcendence beyond the bounds of creation, where he exists in an eternity beyond time and space.

Speculative theology invites us to contemplate the divine purpose behind the creation and recreation of the universe. We can glimpse God's ongoing work throughout the centuries, bringing order out of chaos and life out of emptiness. His grace and mercy extend beyond the limits of time and space, offering hope and redemption to all creation.

The notion of an infinite cycle of creation and recreation leads us to reflect on the promise of the Scriptures of a new heaven and a new earth, where justice will reign forever. This hope transcends the limitations of human understanding and leads us to trust in divine providence, even in the face of the unfathomable mysteries of the universe.

I want to express a conception of a primordial universe as a cosmic womb, pregnant with the new universe we know, is an intriguing metaphor that invites us to contemplate the depth of the creator's work. In this perspective, the ancient universe acts as a receptacle of potentiality, ready to give birth to a new creation through God's intervention. This analogy can be understood in light of contemporary cosmological theories, which suggest the possibility of multiple universes coexisting in a vast multiverse.

As the new universe emerges from the womb of the ancient universe, it expands and evolves over time, following the laws established by God. This process of expansion and evolution can be compared to the big bang theory and the ongoing expansion of the universe, as postulated by modern cosmology. However, like all things in creation, this universe is destined to age and eventually die. This notion of cosmic death echoes scientific predictions about the ultimate fate of the universe, such as heat death or gravitational collapse.

This cycle of cosmic life and death reflects the finitude and transience of all creation, while God, in his infinite wisdom, continues to exercise his sovereign control over the destinies of the universes. This view can be understood within the context of the anthropic principle in cosmology, which postulates that the physical laws of the universe are finely tuned to allow for the existence of life and consciousness.

The return of the new universe to the womb of the ancient universe symbolizes not only a cycle of renewal but also a continuity in God's creative work. Under the same laws that governed its first existence, the new universe undergoes a process of recreation, preparing for a new era of existence and manifestation of divine glory. This view finds support in the biblical promises of new heavens and a new earth, where God's redemptive work will culminate in the complete restoration of all things. Thus, your theory provides a fascinating theological and cosmological perspective on the nature of creation and divine intervention in the universe.

Just as the resurrection of Jesus brought renewal and life after death, the conception of the universe as a cosmic womb invites us to contemplate the possibility of continuous and perpetual rebirth within the cosmos. Just as Jesus emerged from the tomb to a new life, the new universe arises from the womb of the ancient universe, bringing with it the promise of a new era of existence and revelation of God's glory and supernatural.

Similar to Jesus' resurrection representing a victory over death and darkness by rising with strong splendorous light, the emergence of the new universe from the cosmic womb symbolizes a triumphant manifestation of God's creative power over the forces of destruction and chaos. Just as Jesus brought hope and renewal to humanity, the new universe brings with it the promise of a bright future filled with the grace and mercy of God.

In this metaphor, the cycle of cosmic creation and renewal reflects the profound spiritual truth of resurrection, where death is transformed into life, darkness into light, and chaos into order. Just as the resurrection of Jesus brought redemption and salvation to humanity, the continuous creation and recreation of the universe proclaim God's eternal victory over the power of evil and the promise of eternal life for all created beings.

Just as a single sperm, among millions, contains the potential to generate a new life, the primordial point of the big bang can be seen as a cosmic seed containing all the potential for the emergence of the universe as we know it. Just as we do not know what exists before the moment of conception, we have no knowledge of what existed before the primordial point of the big bang. However, just as a tiny sperm gives rise to a complex and diverse life, the primordial point gave rise to a vast and ever-expanding universe.

This theory suggests that, just as sperm fertilizes the egg to initiate the process of developing a new being, the primordial point of the big bang initiated the process of developing the universe. As the universe expanded, evolved, and developed, just as an embryo develops in the mother's womb, conditions were favorable for the formation of stars, galaxies, planets, and eventually life.

This metaphor leads us to reflect on the potential contained within every particle of the universe and how this potential manifests over time. Just as each cell in the human body contains the genetic code for the complexity of life, each particle of the universe contains the fundamental laws of physics that govern its evolution and development.

This theory also invites us to contemplate the interconnectedness of all things in the universe, from the smallest atoms to the largest galaxies. Just as each cell in the human body plays a vital role in the functioning of the organism as a whole, each part of the universe contributes to the harmony and balance of the cosmos.

In the fascinating journey through the world of quantum physics, we are confronted with a series of inexplicable behaviors exhibited by subatomic particles. From the wave-particle duality, which challenges our intuition by presenting particles behaving as waves and vice versa, to Heisenberg's uncertainty principle, which imposes fundamental limitations on our ability to accurately predict the properties of particles. Quantum entanglement leads us to question the very nature of space and time, while quantum superposition challenges our understanding of reality by allowing particles to exist in multiple states simultaneously. The mysterious phenomenon of radioactive decay confronts us with the inherent randomness of unstable particles, making it impossible to predict with certainty when a decay event will occur. These and other enigmas of quantum mechanics invite us to rethink our traditional conceptions of

nature and explore a universe where randomness, probability, and non-
locality become the cornerstones of a new understanding of reality.

The theory of the universe within the universe, or "the thing within
the thing," offers an intriguing perspective on the mysteries of quantum
physics. Considering the possibility of a previous universe that gave birth
to ours, through a process of "dead womb" that conceived the current
cosmos, we are led to speculate about the nature of universes and their
interconnectedness. If there is a universe within a universe, there could
be an infinite sequence of universes, each emerging from the vacuum of
black holes and lifeless dimensions of its predecessor. This theory suggests the existence of all interconnected and unknown laws of physics,
acting in black holes with singularities upon singularities, representing
channels between distinct universes. Black holes, in turn, can be seen
as portals connecting these realities, enabling the transition from one
universe to another.

When we observe the behavior of subatomic particles in the context
of quantum mechanics, many of their strange behaviors can be attributed
to the influence of a previous universe, where the laws of physics were
different. The walls of a cosmic "womb," ruptured by explosions from
this previous universe, may explain inexplicable events in our cosmos,
challenging the known laws of physics and quantum mechanics. The idea
of a "cosmic mirror" between universes cannot be dismissed, suggesting
the existence of unknown dimensions and distinct physical laws. These
premises, although speculative, offer an intriguing explanation for the
behavior of particles in the quantum world, suggesting that the laws of
physics may be more complex and interconnected than we imagine, reflecting the infinite wisdom of God.

The metaphor of the primordial point of the big bang as a cosmic
seed that exploded within a young expanding universe, created by God,
helps us conceive a new proposal for the emergence and evolution of the
universe as a process of continuous growth and development, driven by
the intrinsic potential contained in every particle of the cosmos.

# Chapter V

## The Purpose of God
## in Creating the Cosmos

THE EXPLORATION OF THE theme of creation as an introduction to the kingdom of God is a fascinating approach that offers profound insights into the divine purpose behind the cosmos. It is interesting to observe how the influence of Greek philosophy has shaped Christian theological thought throughout history, especially regarding matter and the notion of the destruction of the cosmos. However, by studying the Scriptures, we can perceive the Judeo-Christian view that the entire cosmos will be recreated by God, in contrast to the Greek belief in the dissipation of the universe.

This biblical understanding leads us to a renewed perspective on the future of the universe, where complete restoration is on the horizon. Instead of a final destruction, we await the arrival of new heavens and a new earth, where redemption will encompass all creation. This view does not promote an escapist theology but encourages us to face the challenges of earthly life with hope and confidence in God's redeeming work.

Paul's perspective on life on this earth, despite sufferings and hardships, resonates deeply with those who have an intimate relationship with God. Just as he preferred to remain on earth because of Christianity, we recognize that our presence here has a divine purpose. Christianity, as the community of believers, is the bond that ties us to the Earth and motivates us to remain committed to God's mission in this world.

This understanding leads us to a deep appreciation of the importance of Christianity in our lives, as it is through it that we experience the presence and purpose of God tangibly. Our family, within the context of Christianity, becomes an integral part of this body of believers who share the journey of faith and support each other along the path of life. Thus, we find meaning and purpose not only in the hope of heaven but also in the responsibility and service to the community of faith here on earth.

Family in the context of Christianity transcends blood ties, as it is composed of those who are united in Christ, sharing spiritual communion and a common purpose in God's work. If the deepest desire to live is not centered on the church, on God's work on earth, it indicates a possible failure in understanding the Christian faith.

We should not fall into the trap of adopting a false spirituality that belittles the material blessings that God grants, such as a good home, a dignified livelihood, or material comfort. However, our deepest desire should be to be in communion with the Lord and to live for his glory, recognizing that all things, including our possessions and our family, belong to God and are meant to reveal his glory.

By understanding that heaven and earth are spiritually connected and that we were created to glorify God, our vision expands to encompass the divine purpose for all creation. The message of creation leads us to recognize that we are part of a cosmic kingdom, where the Earth is also under God's dominion and authority. This allows us to integrate the kingdom of God into our theological understanding in a more comprehensive and profound way.

It is important to correct misconceptions about the kingdom of God, which often limit its application only to the kingdom of heaven, ignoring its presence and influence here on earth. When we understand that God reigns over the entire cosmos, including the Earth, we are challenged to live according to the principles and values of the kingdom of God in all areas of our lives, without alienating ourselves from the world around us.

This balanced understanding frees us from the traps of materialism and asceticism, allowing us to live as citizens of the kingdom of God in this world, while eagerly awaiting the final consummation of the kingdom in the new creation promised by God.

The view that the city, its buildings, and concentrations of people are inherently evil is a mistaken interpretation. Sin resides in the human heart, and it is humanity that corrupts the environment around them, not the other way around. Cities and their physical structures are creations of God and do not have the power to corrupt people.

Stories like those of monks who saw the city as an impure place reflect a distorted mindset that attributes evil to the external environment, when in fact sin lies within people themselves. It is a misconception to believe that the external world, with its cultural and social influences, can contaminate the church. In reality, it is the absence of the life of God in people's hearts that makes them susceptible to corruption by the world.

The gospel offers life, hope, and confidence, enabling us to resist the negative influences of the world. If the life of God dwells in us, we will not be dominated by lifestyles contrary to the principles of the kingdom of God. It is crucial to understand that the Earth and everything in it are not inherently evil but await the manifestation of God's glory through his children.

Therefore, it is essential to abolish the idea that the cosmos and the Earth are inherently evil. On the contrary, the Earth is a reflection of the expectation for the manifestation of the children of God and the revelation of divine glory.

As Christians, we are called to live as witnesses of God's light in this world, transforming it with love, grace, and truth.

The Scriptures teach us that the heavens and the earth reveal the glory of God, manifesting his greatness and majesty. Therefore, it is crucial to understand the distinction between the "world" and the Earth. While the "world" refers to a philosophy or mindset that is evil and opposed to the Scriptures, the Earth itself is a wonder created by God for his glory and for our enjoyment. That is why the church should be concerned with ecological issues, such as the preservation of forests and rivers.

By neglecting these issues, we are, in fact, neglecting the future of our children and future generations. The destruction of forests and natural resources is a manifestation of human corruption and greed, and the church cannot remain silent about it. We must take a firm stand against any illegal activity that degrades God's creation, as it was made for his glory and praise.

It is important for the church to take the lead on these ecological issues, even more than secular organizations like Greenpeace, because we are God's representatives on Earth and should act in accordance with his purposes. When we see the destruction of nature, we cannot remain silent. We must denounce and oppose all forms of harmful ecological exploitation, understanding that God's creation is a testimony to his greatness and love.

Therefore, as we reflect on the theme of creation, we must also consider these ecological issues, as they are intrinsically linked to God's purpose for the Earth and humanity. The biblical account of Genesis leads us to contemplate not only God's initial work but also our responsibility to care for and preserve the world entrusted to us.

The fundamental purpose of God's creation of the cosmos is that Christ reigns over all things. Everything was created for the glory, praise, and greatness of God, and we, as children of God, have been called to reign with him. This is a vision that goes beyond the selfish understanding that the universe was made exclusively for us. We must understand that we are part of a much larger plan, in which everything was created for Christ and for his glory.

It is important to recognize that the message of the church has often been distorted, leading people to adopt a selfish approach to faith. Instead of seeking the glory of God and the fulfillment of his purpose, many end up seeking only to satisfy their own desires and needs. This results in a misunderstanding of the role of the church and our relationship with God.

On the other hand, it is essential to understand that the institution of the church, when properly managed and guided by the Spirit of God, can be a powerful instrument for the manifestation of the kingdom of God on Earth. Although there are cases of mismanagement and corruption, this does not invalidate the importance of the church as a community of believers gathered to worship and serve God.

We must avoid falling into the trap of adopting faith philosophies that do not bring genuine spiritual growth. Often, these approaches lead to disillusionment and drifting away from the true Gospel. Instead, we should seek a deeper understanding of God's purpose for our lives and for creation as a whole, recognizing that we are called to live in accordance with his will and to glorify him in all that we do.

This reflection is profoundly important and reminds us that the enemy can use various tricks to divert people from the true path. Often, we encounter individuals who appear to be involved in religious activities but are actually misguided, worshipping evil instead of God. This underscores the need for spiritual discernment and for us to remain vigilant against the cunning traps of the devil.

It is essential to understand that everything was created for the glory of Christ and for his praise. He is the center of all things, and our lives should reflect that in every aspect. We must rid ourselves of the selfish mindset that places us at the center of the universe and, instead, recognize that we are called to live for the glory of God in all that we do.

We must not fall into the trap of thinking that we have rights to demand from God or to place him in the background. He is sovereign over all things, and we are called to obey his will and his commandments. This is an important reminder that we are all in God's hands, whether we recognize it or not. He is the Creator and Lord of all things, and we must live in reverence and submission to him at all times.

This reflection leads us to understand the greatness and sovereignty of God over all things, including the cosmos and possibly even the multiverse. Whatever the extent of creation, God is above it, infinite in power and majesty.

The idea that the discovery of a multiverse, if ever scientifically proven, could lead us to glorify and worship God even more, as it would reveal even more the vastness and complexity of his creation. This reminds us of how little we know about God and his work and encourages us to worship him for his incomprehensible greatness.

The concept of kingdom encompasses various dimensions, but all are united in a single realm where God reigns supreme over everything. Whether it be the kingdom of heaven, the kingdom of God, or the kingdom of the cosmos, all are under divine authority. When we recognize that the kingdom of God is within us, we understand that we are connected to this supreme realm, participating in the inheritance and co-heirship with Christ. Thus, the kingdom of God within us reflects the cosmic kingdom over which God reigns sovereignly.

This explanation leads us to recognize the ongoing work of the Holy Spirit on Earth, generating life and softening the hardened hearts of sinners. The Holy Spirit dwells within believers, empowering them and guiding them to live in accordance with God's will. The presence of the Holy Spirit is essential so that humanity does not succumb to sin and destruction.

God is sovereign over all creation, from the beginning of time to the present day. He exercises supreme authority over all elements of creation, controlling and governing with wisdom and power. His glory and majesty are manifested throughout the Earth, revealing his creative and sustaining power.

Therefore, recognizing God's sovereignty over the cosmos and over our lives leads us to live in reverence and worship to him. The Holy Spirit continues to act in our lives, empowering us to live according to divine purposes and to proclaim his kingdom throughout the Earth.

This explanation leads us to reflect on the true nature of God's reign and the distorted conceptions we often have about him. God is not an old man sitting on a distant and indifferent throne, nor is he an authoritarian father ready to punish us at any moment.

God's reign is a much broader and deeper concept. He governs not only with authority and dominion over all creation but also with love, grace, and mercy. His reign is not limited to a distant heavenly throne but is manifested in our lives daily through the Holy Spirit who dwells within us.

God is not a tyrant seeking to punish us but rather a loving Father who desires our well-being and guides us with wisdom and care. We must free ourselves from limited and distorted conceptions of him and open ourselves to a true understanding of his love and sovereignty.

Therefore, by recognizing the true reign of God in our lives, we can find peace, security, and joy in his presence, knowing that he loves us and cares for us with an eternal and unconditional love.

It is important to recognize that faith does not exempt us from the difficulties and challenges of life. God does not promise us a life free from trials but strengthens us to face them with courage and confidence in him. When we go through tribulations, we should not question God's love but seek to understand his purposes and learn from the experiences we face.

We must be careful not to fall into the trap of an optimistic faith that expects God to spare us from all suffering. Instead, we should trust in his sovereignty and seek his guidance and strength to navigate difficult times.

It is true that the history of the church is filled with examples of martyrs and saints who faced persecutions and extreme sufferings because of their faith. These examples remind us that the path of discipleship can be arduous but also inspire us to persevere in our journey of faith, confident that God is with us in all circumstances.

Therefore, when facing challenges in our lives, we should remember that God did not promise us an easy life but assured us of his presence and his unwavering love. With faith and confidence in him, we can find the strength to overcome any adversity and grow in our relationship with him.

Yes, it is important to understand that God is present throughout all creation, but this does not mean that he shares the same essence or purpose as other entities or aspects of creation. While God is omnipresent, he remains transcendent and sovereign over everything he has created. The question raised about whether God could be in the heart of Satan is a complex theological issue. While some theological traditions emphasize God's omnipresence and his ability to be present everywhere, others assert that God's presence is associated with his holiness and his willingness to relate to those who seek him in righteousness and faith.

It is important to understand that, although God is present everywhere, his presence is of a different kind from the presence of other creatures. His presence does not imply approval or communion with evil but rather the exercise of his sovereignty and control over all creation. Therefore, while God is present throughout all creation, his presence should not be confused with the identity or nature of other entities or aspects of creation.

It is true that the work of the Holy Spirit is not limited to specific places or certain religious contexts. He can touch and transform lives anywhere and in any situation. There are numerous testimonies of people who have experienced powerful encounters with God in unexpected moments and places.

God is infinitely merciful and gracious, and his presence and power are not restricted to sacred spaces or formal religious practices. He is active throughout the world, seeking to reach willing hearts, regardless of where they are or what they have done in the past.

It is important that we recognize that our relationship with God should not be based solely on receiving blessings or favors from him, but also on offering him our praise, worship, and service. We are called to worship and glorify God not only for the benefits he grants us but simply for who he is: the Creator and Sustainer of all things, worthy of all honor and praise.

The cosmic kingdom is comprehensive yet finite in its extent, unlike the multiverse, which, even if it exists, is also finite. Infinity is exclusive to God, who is eternal in his greatness and majesty. While he is eternal by nature, humans also have an inherent eternity, whether in communion with God or in the lake of fire, according to their acceptance or rejection of Jesus. Understanding this principle is essential to comprehend the purpose and destiny of each individual.

As I contemplate the concept of the lake of fire, I feel a deep unease. It is as if the conscience burns, not only with a physical fire but with the weight of the choices that shaped our lives. I imagine looking across and realizing the magnificence of God's presence, an opportunity we squandered through our own folly. Everything there is imbued with glory, splendor, and we, confined to our own separation.

Often, I have heard the question: how can a just God allow punishment to be infinitely greater than the finite sins we commit? It is a query that resonates in many hearts, including my own. And yet, I pause to ponder on God's infinite love, manifested on the cross, abundantly offered to all. It is as if he extends to us a proposal of love and mercy, which many reject by their own choice.

I understand that the lake of fire is not an eternal punishment imposed by a cruel and sarcastic God. No, I reject that notion. I believe the fire represents the separation from God, the burning awareness of what could have been but was not. And it is precisely because of God's love that he does not desire anyone to suffer eternally. He waits, with infinite patience, for all to come to repentance and find the abundant life he offers.

It is crucial not to omit the message of hell, even if it is uncomfortable. Jesus, in his divine wisdom, spoke about hell more than fifty times. We cannot ignore its reality. It is a place of suffering, of separation, but also a place of missed opportunity. Hell is not the final destination, but the lake of fire will be the end of all human memories forever; hell was proposed in the message of a reconciliation process with God in life.

In the end, I realize that the cosmic kingdom, with its beauty and comprehensiveness, is a manifestation of God's greatness. It extends in light, sky, water, and dry land, expressing life in all its forms and amplitudes. In this kingdom, the love of Christ shines as a light, reminding us that even in the face of the reality of hell, there is hope and redemption available to all who seek the face of the Lord.

In the vast scenery of the cosmic kingdom, we can glimpse different spheres of authority and influence. There is the kingdom of men, where humanity reigns at various levels of government and authority, often granted by God as divine permission to promote peace and social order. In this realm, monarchs have ruled and leaders have governed, recognizing that their authority is granted by God himself.

However, alongside this human kingdom, there exists a darker sphere: the parasitic kingdom, ruled by Satan. This kingdom, though existing by divine permission, is characterized by distorted authority and corrupted governance. Satan, the prince of this realm, reigns over a sphere of influence that relies on human permissiveness and circumstances to subsist. His legality to reign was granted by Adam through the original sin, and with every human failure, he finds an opportunity to exert his power.

It is intriguing to consider that God, in his infinite goodness, could have annihilated Satan completely. However, he chose to allow evil to exist for a determined time, exercising his justice and sovereignty in mysterious ways that surpass our understanding. Although Satan and his followers are beyond divine forgiveness, God continues to love his creation, even those who choose the path of evil. His wrath is an expression of his justice, constantly seeking to call people to repentance and salvation.

Just as God had compassion for Cain, offering him protection and opportunity for redemption, he continues to extend his mercy to all who heed his call. However dark the parasitic kingdom may be, the light of Christ's kingdom shines with hope and salvation for all who seek refuge under his sovereign authority.

While we revel in the love of God, it is essential to recognize the reality of Satan's parasitic kingdom, which manifests in tangible and often inexplicable ways. During my early Christian walk, I witnessed events that defy known physical laws, revealing the vivid and disturbing presence of this realm. I vividly remember a moment when, while conversing with my father, a cup inexplicably spun and shattered before us when speaking of the devil, a phenomenon that defies any scientific explanation. I also witnessed light bulbs exploding without apparent cause and inexplicable events during our prayer times in open places. Many times, I witnessed demonic manifestations where I had to cast out these demons in the name of Jesus, including manifestations in public places full of people like buses.

On other occasions, I experienced more direct encounters with the parasitic realm, including the vision of a draconian figure emerging from a mountain and disappearing underground. These experiences not only confirm the reality of the power Satan wields in this world but also highlight the importance of recognizing and confronting this sphere of spiritual influence.

Once, in a city, a gospel minister, aware that I had the gift from God to cast out demons, invited me to her office to expel a demon that was oppressing and causing suffering to a federal judge, leading to severe depression and suicidal thoughts. As I began the prayer and laid hands on her, a spirit endowed with wisdom and intelligence manifested and threw her to the ground. This demon began to speak, revealing that it had received a substantial offer to corrupt the spirit of that judge and lead her to destruction, resulting in the ruin of her marriage.

This experience made me realize that demons operate in all social spheres, influencing decisions and even using authorities to enact unjust laws in the name of God and legal knowledge, oppressing people's spirits. I understood that Lucifer desires to seize any available social structure to use it for his own purposes and personal pleasures.

It is undeniable that the kingdom of Satan is real and present, but as believers in Christ, we are called to operate in the kingdom of God. Instead of excessively focusing on the enemy's schemes, we should seek God with fervor and dedication. Unfortunately, I often see churches spending more time and energy talking about the devil than proclaiming the power and victory of Christ. We must remember that as followers of Jesus, we are protected by his power and should not fear the evil influence beyond what is allowed by God.

While I have witnessed numerous manifestations of the parasitic kingdom, I remain firmly convinced that those who belong to Christ are shielded by his sovereign authority. The Scriptures are clear in stating that evil cannot touch those who are of God, and this promise is a source of security and comfort for all who trust in him.

As we continue our spiritual journey, may we remember that the church exists to bring the Gospel to all, offering the path of salvation and liberation in Christ. Although I have accumulated a wealth of experiences in the field of spiritual warfare, I recognize that the true mission of Christianity is to proclaim God's redeeming love, rather than engaging in battles that distract and divert our attention from the true purpose of the gospel.

While my approach is generally more theoretical, I cannot ignore the practical and intriguing experiences I have witnessed throughout my spiritual journey. On one memorable occasion, I found myself amidst a crowd gathered around a house where a woman was possessed by a demon. This event was a vivid example of the tangible reality of the spiritual realm and the importance of remaining vigilant in the face of such manifestations.

As I approached the situation, I witnessed the efforts of the priests to cast out the evil spirit, but they were struggling to maintain control of the situation. When I finally stepped forward, the demon acknowledged my presence with a mixture of intimacy and disdain, highlighting the personal and often insulting nature of interactions between humans and the realm of darkness.

In a moment of direct confrontation, I invoked the name of Jesus and commanded the demon to leave, and within seconds, the woman was freed from its oppressive influence. This experience is a powerful reminder of the power of the name of Jesus over the forces of evil and the importance of standing firm in our faith in the face of spiritual challenges.

Throughout my journey, I have also encountered situations where individuals seemed to be under the influence of the Holy Spirit, but were actually possessed by the evil one. It is crucial to develop spiritual discernment to distinguish between genuine manifestations of the Spirit of God and the deceptive imitations of the enemy.

While these experiences are often dramatic and disturbing, they serve as powerful reminders of the reality of conflicting spiritual realms and the need to remain anchored in the truth and power of God. By equipping ourselves with spiritual discernment and a deep understanding of the authority we have in Christ, we can face these challenges with confidence and determination.

Throughout my spiritual journey, I have witnessed countless manifestations of Satan's parasitic kingdom, both within and outside the walls of the church. These vivid experiences highlight the tangible reality of the spiritual battle that is constantly ongoing around us.

I remember a particularly striking incident involving a friend when I was young, Sandro, while we were on a hill praying. Suddenly, a woman began to convulse in a strange manner, and Sandro, with spiritual discernment, alerted me to the situation. In an instant, he approached and, with authority, expelled a legion of demons that possessed that woman. This event serves as a poignant reminder of the need to remain vigilant in the face of manifestations of the kingdom of darkness and to utilize the authority we have in Christ to confront them.

Unfortunately, it is not always easy to discern between genuine manifestations of the Holy Spirit and the enemy's counterfeits. Many times, individuals are deceived and end up being influenced by evil disguised as something divine. I recall an occasion when a sister, who regularly delivered prophecies, was revealed to be possessed by a legion of demons when I prayed for her. These incidents highlight the importance of developing spiritual discernment and remaining steadfast in God's truth.

Satan's parasitic kingdom is not limited only to the church's exterior; unfortunately, it has also infiltrated many congregations. I remember cases where Satanists were active within the church, teaching heretical doctrines and sowing discord. These situations serve as a vivid reminder that we must remain vigilant and be attentive to signs of malignant influence, even within the seemingly safe environment of the church.

However, despite the frightening reality of the parasitic kingdom, we can find comfort in God's sovereignty over all spiritual realms. He rules over all things with justice and fairness, and has granted authority to the church to confront the forces of evil. Every believer has the power and authority in Christ to resist the enemy and stand firm in the truth.

Therefore, as we navigate this world filled with spiritual challenges, we must remain constantly vigilant, armed with spiritual discernment, and trusting in the authority we have in Christ. Only then can we confront the enemy's schemes and stand firm in God's truth and power.

Over the years, I have learned and experienced the authority given to every believer over principalities and powers. This authority is not an exclusive privilege of renowned leaders or prominent figures in the church, but is conferred upon every child of God who invokes the name of Jesus.

It is crucial to understand that this authority does not reside in our own strength or ability but is derived solely from the name of Jesus. When we confront the enemy, it is not by our own merit, but it is the power of the name of Jesus that compels demons to submit and leave. Throughout my nineteen years of ministry, I have never needed to use the name of Jesus more than once to cast out a demon. Once the name of Jesus is invoked, demons must obey and leave immediately.

However, I have observed with concern many ministers unnecessarily prolonging the process of expulsion, spending hours in confrontations with demons. This not only lacks biblical basis but also demonstrates a lack of understanding of the authority we possess in Christ. We should not engage in prolonged debates with the enemy; instead, we should confidently declare the name of Jesus and expect immediate obedience.

The authority we have in Christ is akin to the authority exercised by a police officer or a traffic warden. Just as they do not need to debate with offenders but simply enforce the law, we too should exercise our spiritual authority confidently and without hesitation.

It is important to remember that this authority is not limited only to demons; it extends over all circumstances and challenges we face. Just as we have authority over demons, we also have authority to rule over life situations with the same confidence and faith in the name of Jesus.

Ultimately, we must understand that the authority we have in Christ is not the result of our own efforts but is a gracious gift from God. He has granted us this authority so that we can confidently advance in fulfilling his will and purpose in our lives and in the world around us.

Since I committed to walk with God in Christ, my life has been a constant battle with Satan and his demons. There wouldn't be enough pages to recount the countless cases of demonic manifestations and spiritual experiences that are beyond any scientific understanding I can imagine.

I have witnessed people possessed by demons breaking barbed wire fences with their chests, and others being chained because nothing could hold the demon back; the family saw no other solution to protect their own daughter from death. There was a case where a ten-year-old girl needed over eight women to restrain a child possessed by a legion of demons.

In another abnormal situation, I entered a house to pray, and when I called out the first time, numerous children fell possessed by satanic powers. As rational as I am about issues that require this reasoning, I have lived experiences that are beyond my comprehension.

# Chapter VI

## The Place of Man and Woman in the Cosmos

THE FALL OF HUMANITY only occurred because man was created, and the need for justification arose only because man sinned, disobeying God and violating the covenant established. In the narrative of creation, we can discern God's zeal regarding man and woman, thus understanding the profound extent of his love for us. It is for this reason that I have dedicated so much time to explore this theme in my journey, sharing the Gospel and teaching the Scriptures.

When we understand the purpose for which we were created, our life flows more harmoniously and lightly, relying on the assistance of the Holy Spirit. God created us with a defined purpose, and even before we came into the world, he had already mapped out a plan for our existence. This is intrinsically linked to the idea of election. Before our earthly manifestation, we were already part of divine plans and purposes.

God loved us even before we were here; he desired us and planned us in advance. He created us to operate according to a specific purpose: to praise and glorify him eternally. In conceiving our existence in eternity, he desired us to follow the longing of his heart. Therefore, studying the Scriptures is equivalent to examining the manual that the Creator has bequeathed to guide our earthly journey.

Countless people face frustrations, disappointments, and loss of direction simply because they did not understand the reason for which they were created. The first and primary reason is to praise and glorify God eternally. The second reason why God created us is so that we can enjoy an intimate relationship with him and express his love through our lives. God desires us to live in communion with him, experiencing his presence and enjoying the fullness of life that he offers us. He created us to experience his goodness, his grace, and his mercy daily.

Furthermore, God created us to fulfill a specific purpose in his redemptive work in the world. He called us to be agents of transformation and instruments of his love and justice in this broken world. He equips us with unique gifts and talents so that we can serve others and make a difference in our midst.

Therefore, in addition to praising and glorifying God, we were created to live in deep intimacy with him, expressing his love to the world and fulfilling his purpose for our lives. It is in this journey of relationship with God and service to others that we find true meaning and fulfillment.

God desired in his heart to extend his kingdom, his kingdom project. And the text we will discuss now is precisely the one that describes God placing man and woman in the garden, the place of the creation of man and woman. God created man and woman in his image, according to his likeness. The idea of an image, when we observe an image, is to represent the real figure. Similarly, when God created us, he desired that, when we look at a human being, we see there a genuine representation of his essence.

If we pass through a square and see a statue of Ayrton Senna, we will immediately recognize that figure as Ayrton Senna. This is because a statue represents and symbolizes the original. Thus, we were molded in the image and likeness of God, with the purpose of expressing the glory, greatness, and majesty of the Creator. We are on this earth to glorify the name of the Lord, to manifest the greatness of God, and to appreciate the beauty of the Creator, who formed the heavens and the earth. We are here to exalt the name of Jesus. And the Holy Spirit dwells on earth to assist us in glorifying Jesus, honoring him, loving him, desiring his company, walking by his side, serving him, preaching his word, and spreading his teachings in this world.

God, in creating man and woman, granted them a place in the cosmos. I always question the Lord about the purpose of our existence on this planet. I ask questions that may seem naive, but they reflect a deep intellectual reflection, especially when I observe the numerous conflicts, the unrestrained greed, the usurpation, and the presumption that permeate our society. For example, I observe how people struggle for resources like oil and minerals, dedicating a whole lifetime to acquire a piece of land and build their own residence or to enroll in government housing programs, aiming to secure the right to have a home. These situations lead me to question the divine will behind our earthly circumstances. Could we not have been placed on a planet like Jupiter, adapted for human life, where each couple would have their own country? Given the vastness of the universe, it could have been a viable possibility. However, God chose to place us here, on Earth.

One of the reasons I envision for this divine choice is the need to cultivate unity and mutual love among us. If we were on a planet like Jupiter, adapted for our habitat, it would be rare to find friends and brothers frequently because the distances would be immense. However, God wants us to be close to each other. In more compact environments, such as in a small dwelling, we have the opportunity to see each other constantly, which strengthens the bonds of communion and friendship. Conversely, in larger locations like Jupiter, we could go decades without encountering each other. That is why he placed us on this planet, so that we could express love and unity for one another. Jesus emphasized this message by stating that the world will recognize his disciples by the mutual love they cultivate.

When God placed man and woman in the Garden of Eden, he had various purposes in mind. In addition to multiplying and replenishing the Earth, he entrusted us with governing it, manifesting the greatness and beauty of his kingdom. Our role also involves constituting the family of the covenant, true worshipers who exalt the Creator. Furthermore, we are called to administer the planet's resources and wealth fairly and equitably. Unfortunately, the absence of kingdom teaching and presence has caused numerous social injustices, depriving many of basic necessities such as food, health, housing, and education. However, God has granted us the resources necessary to meet all these needs, and it is up to us, as stewards of this planet, to implement these solutions in a way that reflects his love and justice over all humanity.

There are resources so that no one, neither creatures nor children, suffer. But what prevails is the spirit of corruption, the spirit of greed, of the children of the serpent, of usurpation, and deception, which have deceived even many chosen ones, occupying the hearts of the chosen ones and promoting injustice. So, when we understand the message of creation and God's faithful purpose, things truly take a new turn in our lives. I live by purpose. I walk by purpose. Everything I do is purpose-driven. I don't make any decision, I don't take any action, that is not grounded in God's purpose.

I don't do anything in my life without first seeking God, without first seeking God's heart, God's desire for my life, and based on words that God has spoken to me. From the early years of my conversion, God gave me words about the desire of his heart, about his purpose in my life. I just wait for God's plan to manifest because I understood the purpose. I understood what God wants from my life.

So, I live in tranquility, in peace, and let God's word reign in my life. And you have to walk by purpose. Not by what you see or by what you think, but according to the Scriptures, according to God's purpose and promise for your life, and learn to wait for the Lord's opportune time, when he will make his word come to pass and establish itself in your life, whether in ministry, in missions, in entrepreneurship, in business, in preaching the gospel, or any other area.

You need to wait. When God gave a word to David that he would be king over all Israel, that word took many years to fulfill. David, he walked through caves, was persecuted, was humiliated, suffering in the deserts and fields, waiting for about fifteen years until he was established in the entire nation of Israel. To what extent are you willing to say, "Lord, I believe in You"?

For how long will you be willing to declare, "Lord, I trust in your purpose; I trust that you are faithful and true, and that your word does not fail"? What is happening today? We see a church influenced by immediate societal thinking rather than rooted in Scripture and its propositions. People expect things to happen, but they have already occurred. They won't happen; they have already happened in your life and mine. We are simply on a journey, advancing towards God's purpose. But many people are expecting immediate results because they have not understood that these events unfolded long before the foundation of the world.

There is nothing that God does in my life here that he hasn't already done before. There is nothing that God does in your life today here and now that he hasn't already done before in eternity. You're not here to wait for it to happen; that child you're carrying in your belly, God already conceived that you would have, as a child in eternity. Everything that happens in your life, God has already determined beforehand by his word, his desire. He has already outlined all the purpose, all the plan to establish in your life, and what you need to do is simply listen to the word of God. Simply what will happen with disobedience may take a little longer? I think so, simply God gives the word and the gospel, the Scriptures, for us to understand and walk within his perfect will. But if you don't listen to the word, the voice of God, it may simply take a little longer, but God will fulfill it just as he desired. He won't fail, because God is not a man, that he should lie or repent.

So, you see, God created man and woman and placed them in the cosmos, placed us in the Milky Way, as I mentioned, placed us on planet Earth, in Eden, in the garden. Why do you think this life, sometimes, is so distressing, so painful? It brings you so much suffering, so much anguish, so much sadness at certain times, frustrations, disappointments, conflicts, resistance, pains, hurts, uncontrollable desires often? Because you were not made to live where you're living. I was not made to live in the habitat where I am living. I believe that at no point in God's mind did he plan cities as we conceive them. This, I believe, is a creation of man, of government, of the city, a model based on human ideology.

God has a model of city and government represented in the New Jerusalem. This model differs substantially from the structure we are accustomed to experiencing. Currently, we inhabit urban areas that were not initially planned by the Creator. Originally, God designated for man and woman a home in Eden, a garden full of beauty, life, peace, and harmony. However, the circumstances in which we live today in cities are often marked by turbulence.

It is common to face situations of conflict in traffic, chaos and tensions in the workplace, as well as difficulties in interpersonal relationships, including marital and family problems. These adversities generate anguish and sadness, distancing us from the ideal of peace and harmony originally planned by God.

The world we live in is characterized by wars, conflicts, contentions, divisions, and injustices. However, Jesus offers us a peace that transcends earthly circumstances. He presents us with a genuine peace, distinct from that offered by the world. Therefore, to experience true peace in our hearts, we must seek it in the Prince of Peace, the Lord of Lords.

In the face of daily challenges and dilemmas, it is crucial to understand that we were destined to live in Eden, the place designed by God. The life we currently experience does not correspond to the original purpose of the Creator. However, through humanity's fall, we find ourselves on a journey back to Eden, where we can fully enjoy divine blessings.

Eden was a place characterized by abundance and prosperity. There was no shortage of resources there, as God supplied all needs. However, after the fall, humanity was expelled from this environment of abundance, and nature also suffered the effects of sin, resulting in disorder and conflict.

Despite the devastation caused by sin, we must maintain hope in the complete restoration promised by God. He will lead us back to the place of eternal provision and true peace. The beauty and harmony originally present in Eden will be restored, and we will once again enjoy perfect communion with our Creator.

Therefore, let us understand that the life we currently live does not reflect God's original purpose for humanity. We are on a journey back to Eden, where we will find fullness, peace, and harmony. May we seek true peace in the Prince of Peace, confident in the complete restoration that God has promised us.

When I dump pollutants into a river, I am contributing to the destruction of the environment. By discarding, for example, a tire, which will take years to decompose, into the waters of a river, I am harming not only the present environment but also future generations. However, it is common for humans not to consider such consequences in their actions. Their concern usually falls on the short term, ignoring the impact their actions will have on future generations.

In this sense, it is crucial for the church to adopt an ecological consciousness. By addressing botanical and environmental preservation issues, we are dealing with topics that involve the life and sustainability of the planet. We need to have a deep understanding of creation and the impact of our actions on the environment.

Concern for the future of the planet is not only an environmental issue but also a social and economic one. The destruction of natural resources, such as rivers and forests, will result in serious problems for future generations. Global warming, melting glaciers, and deforestation are just some of the alarming signs of this environmental crisis.

Furthermore, it is important to highlight that negligence regarding economic and financial matters also has its consequences. Although it may seem that such issues are not related to spirituality, a lack of financial resources can generate worries and afflictions that directly affect people's lives and their relationship with God.

Therefore, it is essential for the church to broaden its focus of concern, addressing not only spiritual issues about its own life, but also recognizing that all social, economic, and environmental issues are spiritual in nature from the perspective of creation. Awareness of the importance of environmental preservation and care for natural resources should be recurring themes in our Christian teaching and practice.

What I am experiencing at this moment is of paramount importance, yet spiritual awareness is fundamental to understanding the significance of everything that God has created in our lives. There are artifacts created by humans and left in the cosmos. Although our cars, TVs, and cell phones have their importance, I affirm that water surpasses in relevance all these items combined. It is essential to bear in mind that water is much more vital than any electronic device or motor vehicle. Such awareness is grounded in Scripture, as these are resources provided by God to sustain human life.

One can live without a car, without a cell phone, but it is impossible to survive without the water we drink and the oxygen that trees release into the air. Therefore, it is essential to recognize the importance of these natural resources. As I mentioned earlier, I intend to explore this topic further. In the future, I want to write a work solely on ecology within the purpose of God.

The garden was the designated location for the growth and multiplication of the human race, so that they would rule over the Earth. This was the initial purpose, without sin and without fall. It is mistaken to think that there was no sex before the fall. Conception and procreation were carried out through the ordinary means established by God. Sex is not something depraved or perverted, as long as it aligns with the standards and norms established by the Creator.

The Bible affirms that everything God created was good. Therefore, if God created sex, it is good. He ordained that, through sex, humanity would grow, multiply, and fill the Earth. Intimate knowledge of God occurs through relationship, especially by understanding his thoughts and desires as revealed in the Scriptures.

Deepening in the word of God, meditating on it day and night, and through prayer, allows us to develop this intimacy. Prayer is not just speaking, but also listening to what God has to say. Throughout my spiritual journey, I have learned to listen to the voice of God and to seek his guidance in all circumstances.

If a butterfly lands on my window in the morning, or even a sparrow, depending on the message I wish to receive from God, I can decipher it nowadays. Do not just wait for the intervention of a prophet. Animals also bring God's message, observe. Many times, they act as they did with Balaam's donkey. Do you doubt? God used it as an instrument to communicate a message. What did the ravens bring to Elijah? A message that God is good and that he provides for the needs of his servants by bringing them food.

God has various messengers and often sends even angels. Personally, I have been visited by angels on several occasions. Many people think that God communicates only through more conventional means, such as prophets and preaching, which are the main channels. However, there are many ways in which God speaks to us. He can use your spouse, a child, the wind, among others. Just as it happened with Elijah, God employs wind, storm, fire, and various means to make us understand his message. He can speak to us directly, through dreams, revelations, worship, and even through YouTube, where many men of God minister his word.

God can also use books, teachings, and doctrines to communicate his will to us. There is a unique dynamic in the way he speaks to us, and this develops through a prayer relationship. Sometimes we focus too much on expecting to receive a divine message only through a prophet or preaching, but all these forms of communication are supported by the Word of God.

Regarding the issue of stones crying out, this is not just a poetic metaphor, but a literal reality. God can use any means to convey his message to us, including stones. Just as Luther claimed that trees are preaching daily, understanding creation is enough to understand divine communication through it.

The garden was the place intended for growth, multiplication, and intimate knowledge of God. Before the fall, there was no need for intermediation to worship God. Jesus and the angels were present in the garden, communicating the wonders of the Creator. Today, God is even more present, for the Holy Spirit is with us daily, in contrast to the occasional visit he made to our ancestors. Our God is not a God of visits, but a God who is with us constantly, in every moment of the day.

God is present in your life every day, Dona Ehei, Jane, Mila, and Varsilva. You walk with God, and he walks with you, Regina. Even when you are not praising Him, he is by your side. There is no need to cry out, "Lord, come visit us!" He is already there. Are you waiting for a visit from

God? He is with us constantly, until the end of time. Allow me to scandalize you a little more.

The garden was conceived as a place of continuous learning, in relation to the totality of the cosmos. When God placed man and woman in the cosmos, they had an undoubtedly deeper understanding of God's wonderful creation. They dwelled in a spiritual dimension where these things were fully understood. With the fall, science has discovered, in fact, what our ancestors already knew before the fall: that they possessed a deeper knowledge of the material world. This explains why Jesus, in his earthly incarnation, could perform healings, for he understood the nature of diseases and had the power to manipulate them for healing.

Adam and Eve were created to continuously grow in a learning relationship with the Lord. That's why the Lord would visit them to converse with them. The same happens with us. When we deny or neglect knowledge, we are doomed to destruction, as the Scriptures say.

The garden was created with the purpose of expressing the beauty of life and the commitment to covenantal ordinances. There, a covenant was established between God, man, and woman, with the purpose of expressing the beauty of life and fulfilling these covenants. The first mandate was the social mandate, which advocates for good relationships between people, regardless of their differences. We are not at war with anyone but are preaching the gospel, which sometimes causes division. However, we have the social mandate to relate well to all people, respecting them as creatures of God. Although we may disagree with certain behaviors, we should express our position based on the word of God, without promoting campaigns or hostile attitudes against certain groups.

As the Scriptures affirm, we should seek peace with everyone, if possible. However, we know that this will not always be possible. The conception that the church is at war with society is mistaken. We are fighting against values and ideas that are being introduced, even within the church. We must respect people and understand the limits of individual freedom established by law. Humanity should seek peace with each other, despite cultural and ideological differences.

We are here on earth to bring the gospel of God, which offers peace, life, and hope. My desire as a minister of God is that everyone knows this salvation and understands that there is a better life in Christ Jesus, which goes beyond the fleeting pleasures of this world. We must bring the love of God to all who believe, regardless of their condition or life history.

If I do not believe in this inclusive and transformative gospel, I will cease to preach it. I believe in a gospel that saves everyone, from the most virtuous to the most sinful. There is no partiality in this gospel because all those who understand and accept this message are saved. This is our conviction in election, for by election, no one can resist the grace of God.

Christianity fails when it does not stand up for minority groups, when it does not actively combat racism and discrimination. It fails by neglecting widows, the poor, the homeless, and those struggling to obtain daily bread. Furthermore, it fails by presuming that some people will not have a place in the kingdom of God before the consummation of all things. Failure is evident when Christianity distances itself from environmental, economic, and political issues within a nation. Any form of prejudice or discrimination that Christianity allows is a greater sin than the morality it professes in God. This set of failures reveals a disconnect between the fundamental principles of Christianity and its practical application in contemporary society. To achieve true integrity and coherence, Christianity must embrace a stance of inclusion, social justice, and environmental responsibility, aligned with the teachings of love for neighbor and care for divine creation.

# Chapter VII

## Social Mandate for Man and Woman

IN THE EDUCATIONAL CONTEXT, it is essential to recognize the importance of Judaism as a starting point for understanding the Christian faith. The Bible, the sacred book of Christians, is fundamentally a Jewish book. Jesus Christ, the central figure of Christianity, was Jewish, as were the apostles who followed him. The entire biblical narrative portrays the journey of the Jewish people, chosen by God to be guardians of his oracles and to reveal his will to humanity.

At the same time, it is crucial to remember that human history traces back to Noah and his three sons, who spread across the Earth. In this sense, there is no greater degree of love for Jews compared to other peoples, but it is undeniable that they hold a special place in the history of faith and divine revelation.

Understanding Judaism is fundamental for mature and solid Christian faith. The church, by neglecting this aspect, risks becoming immature in faith and falling into theological liberalism. It is through the deep teaching of sacred texts, both from the Old and New Testaments, that the church can achieve spiritual maturity and fully understand God's will.

Therefore, it is necessary to dedicate oneself to theological study and teaching in churches, prioritizing the understanding of Jewish tradition and its relevance to the Christian faith. Knowledge of these foundations strengthens Christian identity and provides a solid foundation for faith practice.

Through understanding the history of Judaism and its interaction with Christianity, believers can deepen their connection with God and live according to his will. Commitment to theological teaching, grounded in Jewish biblical tradition, is essential for the spiritual growth and maturity of the church.

In developing this chapter, we delve deeply into the cultural mandate, one of the fundamental aspects of Reformed theology that is often overlooked. We explore the meaning of the divine mandate present in Genesis, chapter 1, verse 28, which calls humanity to "be fruitful and multiply, fill the earth and subdue it."

This mandate goes beyond simple procreation, being a commission for humanity to exercise its sovereignty over God's creation. The original purpose was for the Eden, the garden planted by the Lord, to extend throughout the Earth, transforming it into a place of abundance, harmony, and communion with the Creator.

As we study the creation mandates, we better understand our responsibilities beyond the church doors. The cultural mandate calls us to create cultures and build civilizations that reflect the values of the kingdom of God. Therefore, the church has the vital role of establishing the kingdom of God in all spheres of society.

The kingdom of God is not limited to spiritual matters but also seeks the redemption of cultures. This means that we should bring the truth of the gospel to every aspect of human life, whether in music, art, education, or even the financial system. The kingdom of God brings social transformation, combating corruption, and restoring integrity in all areas of society.

Historically, we see powerful examples of how the gospel impacted society, whether during the early days of Christianity in the Roman Empire or during the Protestant Reformation. In both cases, economic and social issues were addressed, resulting in profound social transformation.

Therefore, as bearers of the kingdom of God, we are called to be agents of change and redemption in all cultures and societies where we live.

This chapter seeks not only to inform but to inspire readers to understand their role in the redemption of cultures and in building the kingdom of God in all areas of human life.

This is an invitation for the reader to become a catalyst for cultural transformation, bringing the light of the gospel to every corner of society, wherever they may be.

In the current context, we confront the cultural challenges plaguing contemporary society, from political corruption to moral decay, and recognize the urgency of a proactive response from the church. In light of this, we reaffirm the cultural mandate conferred by God upon humanity, calling us to the responsibility of seeking redemption and transformation in all spheres of culture and society.

Reflecting on the historical impact of Christianity throughout history, from its origins to the most recent revivals, we highlight how the message of the gospel has been the catalyst for profound social changes in different times and contexts. In this regard, we emphasize the importance of creative intelligence in cultural redemption, highlighting the urgent need for committed Christian professionals in various areas of society, such as healthcare, education, art, and music.

We need a commitment to God's mission as the way to truly impact the world around us, transforming societies and manifesting the kingdom of God in all spheres of life. This chapter serves as a journey of deeper understanding of the importance of cultural redemption and its potential power to promote lasting social transformation, anchored in the values of the kingdom of God.

As we delve into fertile ground for reflections, inspired by the profound symbolism of the film *The Village*, this feature film, despite its apparent simplicity, offers a rich tapestry to examine the dynamics between the ecclesiastical community and the outside world.

In the plot of the film, we witness an isolated community, built to escape the hardships and evils of the outside world. In response to the problems and conflicts of modern society, the villagers decide to create a safe enclave, shielded from external influences. This gesture, initially motivated by a quest for security and purity, eventually reveals itself as an escape from the challenge of engaging with the broader issues of society.

Similarly, we observe in the life of the church a similar tendency to withdraw from the world around it. Fear of sin and corruption often leads Christians to isolate themselves in their own communities, seeking to maintain a supposed holiness by distancing themselves from the realities of the world. However, as we learn from the narrative of *The Village*, this segregation can not only be unsustainable but also counterproductive to the broader purpose of the church.

Instead of isolating ourselves, the church is called to be a light in the world, actively engaging with the issues and challenges of the society in which it is embedded. As the film vividly reminds us, true depravity does not only reside in the external world but also in the hearts and communities that seek holiness. Sin is not an external force to the church but something that needs to be confronted and redeemed within itself.

Thus, we are invited to abandon the "village myth" that suggests we can fully shield ourselves from external influences. Instead, we should embrace the challenging mission of being transformers in our midst, bringing the light and love of Christ to all the places where we are present.

Considering the example of the protagonist of the movie, who dares to challenge the limits imposed by the village and ventures into the outside world in search of help and redemption, we are encouraged to also step out of our comfort zones and actively engage with the world around us. After all, it is only by facing challenges and embracing the complexities of everyday life that we can truly fulfill Christ's call to be salt and light in the world.

As we conclude this book, it is essential to emphasize the importance of seeking God in all aspects of life. However, this spiritual quest cannot be limited to contemplation and worship within the church walls. It is imperative to understand that part of the divine mandate is the practical application of faith in all spheres of existence.

It is futile to fervently dedicate ourselves to worship if, at the same time, we neglect our role and relevance in academic, professional, and social environments. It is necessary to go beyond, it is necessary to act. It is necessary to bring redemption to the places where God has placed us.

Worship, devotion, and instruction are fundamental, but it is also crucial to take the received message beyond the limits of the ecclesiastical space. Just as the Ark of the Covenant could not be carried by just one person due to its weight, we also need men and women of God to invoke, bring, and carry divine presence.

The gathering and communion of believers are essential to seek the presence of God and take his message to the outside world, thus fulfilling the cultural mandate. This mandate transcends the church walls and extends to all spheres of society, including art, music, sports, medicine, law, and all other areas of endeavor.

Cultural redemption implies being a transformative agent in all areas of life, bringing the light and love of Christ to every place where we are present. It is through our actions, testimony, and influence that we can truly impact the world around us.

Therefore, I conclude this book by emphasizing the urgent need to be not only worshippers but also agents of transformation in all spheres of society, filled with the Holy Spirit and committed to the mission of taking God's message to every corner of the world. May we understand and embrace the purpose for which we have been called, recognizing that we were not made to live in seclusion, but to shine as lights in the midst of the darkness of the world. May this book inspire and encourage every reader to live a life of impact and significance, reflecting the love and grace of God in all they do. Amen.

Throughout the pages of this book, I guide readers on a journey of exploration and discovery in the vast field of astrophysics, where I present a new theory about the universe. This theory does not emerge in isolation, but is the result of a careful immersion in the works of various authors whose perspectives and approaches have deeply enriched my understanding of cosmic mysteries.

Prominent authors such as Lawrence Krauss, Ellis, Carroll, Hawking, and Penrose, among other leaders in astrophysics, have provided valuable insights that have shaped my view of the universe. Their theories about the nature of time and the origin of the cosmos, have served as sources of challenge and inspiration for the development of my own line of thought.

While the theory I present represents an original approach, I recognize the importance of considering and understanding the various perspectives offered by these renowned astrophysicists. Intellectual dialogue and collaboration among thinkers from different disciplines have been crucial in shaping this work.

This book not only represents a continuous effort to unravel the secrets of the universe, but also celebrates the diversity and depth of human knowledge. I hope it inspires readers to explore the wonders of the cosmos and to contemplate the depths of the unknown, just as I have done throughout this journey of cosmic discovery.

# Bibliography

Brueggemann, Walter. *Theology of the Old Testament: Testimony, Dispute, Advocacy.* Minneapolis, MN: Fortress, 2005.

Carroll, Sean M. *From Eternity to Here: The Quest for the Ultimate Theory of Time.* New York: Dutton, 2010.

Carson, D. A., et al. *New Testament Biblical Theology: The Unfolding of the Old Testament in the New.* Grand Rapids, MI: Baker Academic, 2011.

Childs, Brevard S. *Biblical Theology in Crisis.* Louisville, KY: Westminster John Knox, 1970.

Condell, Patrick. *Godless and Free.* Self-published, 2012.

Dempster, Stephen G. *Dominion and Dynasty: A Theology of the Hebrew Bible.* Downers Grove, IL: InterVarsity, 2003.

Einstein, Albert. *The Born-Einstein Letters: Correspondence between Albert Einstein and Max and Hedwig Born from 1916–1955.* London: Macmillan, 1971.

Ellis, George F. R. "Physics and the real world: from the Big Bang to quantum resurrection." *Nature Physics* 1.1 (2005) 2–5.

Goldingay, John. *Old Testament Theology: Israel's Gospel.* Downers Grove, IL: InterVarsity, 2003.

Guth, Alan H. "Eternal inflation and its implications." *Journal of Physics A: Mathematical and Theoretical* 40.25 (2007) 6811.

Hamilton, James M. *God's Glory in Salvation through Judgment: A Biblical Theology.* Wheaton, IL: Crossway, 2010.

Hawking, Stephen, and Leonard Mlodinow. *The Grand Design.* London: Bantam, 2012.

Hebrew Version: Biblia Hebraica Stuttgartensia. Stuttgart, Germany: Deutsche Bibelgesellschaft, 1997.

Kaiser, Walter C., Jr. *Toward an Old Testament Theology.* Grand Rapids, MI: Zondervan, 1978.

Linde, Andrei. "Inflation and the Multiverse." *Reports on Progress in Physics* 80.2 (2017) 022001.

*New International Version (NIV): NIV Study Bible.* Grand Rapids, MI: Zondervan, 2011.

Packer, J. I., et al. *New Dictionary of Biblical Theology.* Downers Grove, IL: InterVarsity, 2001.

Penrose, Roger. *Cycles of Time: An Extraordinary New View of the Universe.* New York: Vintage, 2011.

Ray, Darrel. *The God Virus: How Religion Infects Our Lives and Culture*. IPC, 2009.

Turok, Neil. *The Universe Within: From Quantum to Cosmos*. House of Anansi, 2012.

Smolin, Lee. *The Life of the Cosmos*. Oxford: Oxford University Press, 1999.

Steinhardt, Paul J., and Neil Turok. "Endless universe: Beyond the Big Bang." *American Scientist* 92.1 (2004) 38–45.

Vanhoozer, Kevin J., et al. *Dictionary for Theological Interpretation of the Bible*. Grand Rapids, MI: Baker Academic, 2005.

Vos, Geerhardus. *Biblical Theology: Old and New Testaments*. Carlisle, PA: Banner of Truth, 1975.

www.ingramcontent.com/pod-product-compliance
Lightning Source LLC
Chambersburg PA
CBHW060338100426
42812CB00003B/1037